This series aims to report new developments in physical research and teaching – quickly, informally, and at a high level. The type of material considered for publication includes:

1. Preliminary drafts of original papers and monographs

2. Lectures on a new field, or presenting a new angle on a classical field

3. collections of seminar papers

4. Reports of meetings

Texts which are out of print but still in demand may also be considered if they fall within these categories.

The timeliness of a manuscript is more important than its form, which may be unfinished or tentative. Thus, in some instances, proofs may be merely outlined and results presented which have been or will later be published elsewhere.

Publication of *Lecture Notes* is intended as a service to the international physical community, in that a commercial publisher, Springer-Verlag, can offer a wider distribution to documents which would otherwise have a restricted readership. Once published and copyrighted, they can be documented in the scientific libraries.

Manuscripts

Manuscripts are reproduced by a photographic process; they must therefore be typed with extreme care. Symbols not on the typewriter should be inserted by hand in indelible black ink. Corrections to the typescript should be made by sticking the amended text over the old one, or by obliterating errors with white correcting fluid. The figures (in the original size) ready for reproduction should be inserted into the text. Should the text, or any part of it, have to be retyped, the author will be reimbursed upon publication of the volume. Authors receive 50 free copies.

The typescript is reduced slightly in size during reproduction, therefore a large size of type should be used; best results will not be obtained unless the text on any one page is kept within the overall limit of 18 x 26.5 cm (7 x 10½ inches). The publishers will be pleased to supply on request special stationery with the typing area outlined.

Manuscripts in English, German or French should be sent to Springer-Verlag, 6900 Heidelberg, Postfach 1780.

Die *„Lecture Notes“* sollen rasch und informell, aber auf hohem Niveau, über neue Entwicklungen in der Physik berichten. Zur Veröffentlichung kommen:

1. Vorläufige Fassungen von Originalarbeiten und Monographien.

2. Spezielle Vorlesungen über ein neues Gebiet oder ein klassisches Gebiet in neuer Betrachtungsweise.

3. Seminarausarbeitungen.

4. Vorträge von Tagungen.

Ferner kommen auch ältere vergriffene spezielle Vorlesungen, Seminare und Berichte in Frage, wenn nach ihnen eine anhaltende Nachfrage besteht.

Die Beiträge dürfen im Interesse einer größeren Aktualität durchaus den Charakter des Unfertigen und Vorläufigen haben. Sie brauchen Beweise unter Umständen nur zu skizzieren und dürfen auch Ergebnisse enthalten, die in ähnlicher Form schon erschienen sind oder später erscheinen sollen.

Die Herausgabe der *„Lecture Notes“* Serie durch den Springer-Verlag stellt eine Dienstleistung an die physikalischen Institute dar, indem der Springer-Verlag für ausreichende Lagerhaltung sorgt und einen großen internationalen Kreis von Interessenten erfassen kann. Durch Anzeigen in Fachzeitschriften, Aufnahme in Kataloge und durch Anmeldung zum Copyright sowie durch die Versendung von Besprechungsexemplaren wird eine lückenlose Dokumentation in den wissenschaftlichen Bibliotheken ermöglicht.

Lecture Notes in Physics

Edited by J. Ehlers, Austin, K. Hepp, Zürich and
H. A. Weidenmüller, Heidelberg
Managing Editor: W. Beiglböck, Heidelberg

3

André Martin
CERN, Genève

Scattering Theory: Unitarity, Analyticity and Crossing

Notes taken by R. Schrader, Zürich

Springer-Verlag
Berlin Heidelberg GmbH 1969

ISBN 978-3-540-04641-7 ISBN 978-3-540-36168-8 (eBook)
DOI 10.1007/978-3-540-36168-8

Originally published by Springer-Verlag Berlin Heidelberg New York in 1969.

Library of Congress Catalog Card Number 70-106192
Title No. 3322

Foreword

These lecture notes are based on a course that I gave at the Swiss Federal Institute of Technology during the Summer Semester 1969 at the invitation of Professors M. Fierz, K. Hepp and R. Jost. I am extremely grateful to them for this opportunity they gave me to teach on my favourite subject. Dr. R. Schrader has been kind enough to accept to take notes and reconstruct a coherent version of what I said. If the reader is dissatisfied with the presentation of the material contained in the lecture notes I am to blame because I think that Dr. Schrader did a very good job in disentangling my sineous progression and restating in a more correct mathematical language what I said. Dr. Schrader and I are also very much indebted to Dr. C. Chandler and Prof. K. Hepp who critically read the manuscript and eliminated the most obvious grammatical mistakes. However, they should not be held responsible for the general Un-English appearence of lecture notes originating from a course given in French to a German speaking audience. Many thanks are also due to Miss Hintermann who had the task to type these notes.

André Martin

<u>Contents</u>

I. Introduction and historical background

These lectures will describe the interplay of analyticity and unitarity of the S-matrix. Originally unitarity was used together with some other ingredients to control the magnitude of the scattering amplitude. The first published example of this is the remark by T.D. Lee (CERN report 61-30, p. 63) that the calculation by lower order Fermi theory of the $e\bar{\nu} \rightarrow e\bar{\nu}$ cross section is certainly incorrect because it exceeds $4\pi k^{-2}$ though the only angular momentum contributing is $J = 1$. The first application of such kind of consideration to strong interactions was by Froissart [F 1]. Starting from the postulate of Mandelstam representation, he was able, with the help of the unitarity condition, to obtain an upper bound $\sigma_{tot} < (\log E_{CM})^2$ on the total cross section. Later many other bounds were derived: Greenberg and Low [G 4] obtained the result $\sigma_{tot} < (E_{CM} \log E_{CM})^2$ from field theory, some unnecessary assumptions in the proof of the Froissart bound were eliminated (Martin 1962 [M 5]); and upper bounds for high energy and lower bounds for fixed angle amplitude were obtained in the Mandelstam framework (Cerulus-Martin [C 1], Kinoshita-Loeffel-Martin [K 3]). In addition to these asymptotic bounds rigorous limits on the pion-pion amplitude were obtained for finite values of the arguments by Martin [M 8], for instance, with the standard notations,

$$\left| F\left(\tfrac{4}{3}\mu^2, \tfrac{4}{3}\mu^2, \tfrac{4}{3}\mu^2\right)\right| < 100$$

and sum rule inequalities on the pion pion total cross sections were proved. Another use of unitarity is to allow the enlargement of the analyticity domain of scattering amplitudes. The first example of this is given by the work of Mandelstam [M 3], who was able to enlarge the analyticity domain of a scalar amplitude $\sigma\sigma \rightarrow \sigma\sigma$ by using elastic unitarity. The second example on which we shall spend a lot of time is the application of the positivity properties of the absorptive part to get dispersion relations for fixed positive (unphysical) transfer [M 9], [M 10].

In these lectures we shall not follow the historical order. After some preliminaries concerning three most striking aspects of unitarity: Control of the magnitude of the amplitudes, positivity, and elastic unitarity, we shall discuss the extension of the analyticity domain obtained from field theory with the help of positivity. Later, once the new domain is obtained, we shall get bounds on the scattering amplitude.

II. Notations

In our present discussion we will consider only the elastic two-body scattering amplitude of spin-zero massive, stable particles:

(1) $$A + B \rightarrow A + B .$$

Let k be the momentum in the center of mass system, Θ the scattering angle. Then we introduce the relativistic invariants

$$s = \left((M_A^2 + k^2)^{\frac{1}{2}} + (M_B^2 + k^2)^{\frac{1}{2}} \right)^2$$
$$t = -2k^2 (1 - \cos\Theta)$$
$$u = 2M_A^2 + 2M_B^2 - s - t .$$

It will be convenient to include the process

(2) $$A + \bar{B} \rightarrow A + \bar{B}$$

into our consideration. The total energy is given by $s^{\frac{1}{2}}$ in process (1) and by $u^{\frac{1}{2}}$ in process (2). Denoting the initial momenta by q_1 and q_2 and the final momenta by p_1 and p_2, we have

$$s = (p_1 + p_2)^2 = (q_1 + q_2)^2 ;$$
$$t = (p_1 - q_1)^2 = (p_2 - q_2)^2 ; \qquad p_1 + p_2 = q_1 + q_2 .$$
$$u = (p_1 - q_2)^2 = (p_2 - q_1)^2 ;$$

III a) General Considerations on unitarity, boundedness and positivity

The scattering amplitude $\mathbb{T}$ is defined by the $\mathbb{S}$-matrix as

$$\mathbb{T} = -\frac{1}{2i} (\mathbb{S} - \mathbb{1})$$

such that unitarity gives

$$\frac{1}{2i} (\mathbb{T} - \mathbb{T}^{+}) = \mathbb{T}^{+} \mathbb{T} = \mathbb{T} \mathbb{T}^{+} .$$

The corresponding matrix elements of the elastic two-body processes are described by a distribution [H 1]

$$\mathbb{F}(s, \cos\Theta) \in \mathcal{S}' \left([(M_A + M_B)^2, \infty) \times [-1, +1] \right)$$

If we write

$$\text{III (1)} \qquad F(s, \cos\Theta) = \frac{s^{\frac{1}{2}}}{2k} \sum_{\ell=0}^{\infty} (2\ell+1) f_\ell(s) P_\ell(\cos\Theta)$$

then

$$f_\ell(s) = e^{i\delta_\ell(s)} \sin\delta_\ell(s),$$

where $\delta_\ell(s)$ is real below the first inelastic threshold and generally $\operatorname{Im}\delta_\ell(s) \geq 0$.

Also the differential cross-section is easily written in terms of F :

$$\text{III (2)} \qquad \frac{d\sigma_{el}}{d\Omega} = \frac{4|F|^2}{s} .$$

Let $\underset{\sim}{\Omega}$ represent points on the unit sphere Σ^2 . Then the unitarity of S gives the following relation for $F(s, \cos\Theta)$

$$\text{III (3)} \qquad \frac{2k}{s^{\frac{1}{2}}} \int \overline{h(\underset{\sim}{\Omega}_1)}\, F^*(s, \underset{\sim}{\Omega}_1 \cdot \underset{\sim}{\Omega}_3)\, F(s, \underset{\sim}{\Omega}_3 \cdot \underset{\sim}{\Omega}_2)\, h(\underset{\sim}{\Omega}_2) \prod_{i=1}^{3} \frac{d\Omega_i}{4\pi}$$

$$\leq \int \overline{h(\underset{\sim}{\Omega}_1)} \operatorname{Im} F(s, \underset{\sim}{\Omega}_1 \cdot \underset{\sim}{\Omega}_2)\, h(\underset{\sim}{\Omega}_2) \frac{d\Omega_1}{4\pi} \frac{d\Omega_2}{4\pi}$$

for all square-integrable functions h on the sphere and with equality below the first inelastic threshold. Equation III (3) may also be written in terms of the functions $f_\ell(s)$:

$$\text{III (3')} \qquad |f_\ell(s)|^2 \leq \operatorname{Im} f_\ell(s) \leq 1,$$

the first inequality again being an equality below the first inelastic threshold. Strictly speaking equation III (3') is meant in the distribution sense with respect to s . Then equation III (3') shows that each $\operatorname{Im} f_\ell(s)$ is a positive bounded measure in s. [S2] .

Equation III (3') is a boundedness condition that prevents cross-sections from getting too large and is the ultimate source of all high energy bounds on the scattering amplitude. The question whether $\operatorname{Im} F(s, \cos\Theta)$ is similarly bounded, however, cannot be answered until we know how many partial waves effectively are contributing to the scattering. But this number is not known a priori. Thus by itself equation III (3') is not sufficient to give a bound and extra ingredients must be added in such a way that the number of contributing partial waves is limited. These questions will be discussed in the following chapters. We can say something now, however, if we are willing to accept some uncertainty on the angle at which the bound holds. The following argument is due to Glaser [G 2]:

Let $g(x)$ be a square integrable function on $[-1, 1]$ and set for an arbitrary $\underset{\sim}{\Omega}_1 \in \Sigma^2$

$$g_{\underset{\sim}{\Omega}_1}(\underset{\sim}{\Omega}) = g(\underset{\sim}{\Omega}_1 \cdot \underset{\sim}{\Omega}) .$$

Then

$$\| g \|^2 \underset{\text{def}}{=} \frac{1}{4\pi} \int | g_{\underset{\sim}{\Omega}_1}(\underset{\sim}{\Omega}) |^2 d\Omega = \frac{1}{2} \int_{-1}^{+1} | g(x) |^2 dx$$

does not depend on $\underset{\sim}{\Omega}_1$. In case where $g(x)$ has support only in a neighborhood of $x = 1$, we can imagine $g_{\underset{\sim}{\Omega}_1}$ as a wave packet centered around $\underset{\sim}{\Omega}_1$. A partial wave decomposition

$$g_{\underset{\sim}{\Omega}_1}(\underset{\sim}{\Omega}) = \sum_{\ell=0}^{\infty} (2\ell+1) \, g_\ell \, P_\ell(\underset{\sim}{\Omega}_1 \cdot \underset{\sim}{\Omega})$$

gives

$$\text{III (4)} \qquad \| g \|^2 = \sum_{\ell=0}^{\infty} (2\ell+1) \, | g_\ell |^2$$

and

$$\text{III (5)} \qquad \operatorname{Im} F_{av}(s, \underset{\sim}{\Omega}_1 \cdot \underset{\sim}{\Omega}_2) \underset{\text{def}}{=} \int \overline{g_{\underset{\sim}{\Omega}_1}(\underset{\sim}{\Omega}')} \operatorname{Im} F(s, \underset{\sim}{\Omega}' \cdot \underset{\sim}{\Omega}) \, g_{\underset{\sim}{\Omega}_2}(\underset{\sim}{\Omega}) \frac{d\Omega \, d\Omega'}{4\pi \; 4\pi}$$

$$= \frac{s^{\frac{1}{2}}}{2k} \sum_{\ell=0}^{\infty} (2\ell+1) \operatorname{Im} f_\ell \cdot | g_\ell |^2 .$$

To prove III (4) and III (5) we write P_ℓ in terms of the spherical harmonics

$$P_\ell(\underset{\sim}{\Omega} \cdot \underset{\sim}{\Omega}') = \frac{4\pi}{(2\ell+1)} \sum_{m=-\ell}^{m=\ell} Y_\ell^m(\underset{\sim}{\Omega}) \, \overline{Y_\ell^m(\underset{\sim}{\Omega}')}$$

and then use the orthogonality relations of the Y_ℓ^m on the sphere. Equations III (3'), III (4) and III (5) combined give

III (6) $$\left| \operatorname{Im} F_{av} (s, \underset{\sim}{\Omega}_1 \cdot \underset{\sim}{\Omega}_2) \right| \leq \frac{s^{\frac{1}{2}}}{2k} \, \| g \|^2 .$$

Suppose now that $\operatorname{Im} F(s, \cos\Theta)$ for fixed s is continuous in Θ and that g is chosen to be positive and continuous. The mean value theorem is then applicable:

III (7) $$\operatorname{Im} F_{av} (s, \underset{\sim}{\Omega}_1 \cdot \underset{\sim}{\Omega}_2) = \operatorname{Im} F(s, \cos\bar{\Theta}(s)) \left(\frac{1}{2} \int_{-1}^{+1} g(x)\, dx \right)^2 .$$

Combined with III (6) we obtain

III (8) $$\left| \operatorname{Im} F(s, \cos\bar{\Theta}(s)) \right| \leq \frac{s^{\frac{1}{2}}}{k} \frac{\int_{-1}^{+1} |g(x)|^2 dx}{\left(\int_{-1}^{+1} g(x)\, dx \right)^2} .$$

Choosing for $g(\cos\Theta)$ a function approximating the function which is 1 for $0 \leq \Theta \leq \varepsilon$ and zero else, we have

III (9) $$-2\varepsilon + \Theta_{12} \leq \bar{\Theta}(s) \leq \Theta_{12} + 2\varepsilon$$

$$\cos\Theta_{12} = \underset{\sim}{\Omega}_1 \cdot \underset{\sim}{\Omega}_2$$

Then equation III (8) yields the estimate

III (10) $$\left| \operatorname{Im} F(s, \cos\bar{\Theta}(s)) \right| \leq \frac{s^{\frac{1}{2}}}{k} \frac{1}{1 - \cos\varepsilon} \approx \frac{2 s^{\frac{1}{2}}}{k \varepsilon^2} .$$

Noting that $\left| \frac{2k}{s^{\frac{1}{2}}} \int \operatorname{Im} F(s, \cos\Theta_{12}) \frac{d\Omega_2}{4\pi} \right| = \left| \operatorname{Im} f_0(s) \right| \leq 1$ and writing this integral as $\operatorname{Im} F(s, \cos\bar{\Theta}_{12})$ with $0 \leq \bar{\Theta}_{12} \leq \pi$ we see that III (10) is a generalization of the partial-wave unitarity relation for $\ell = 0$.

We remark that the same estimate for $\operatorname{Re} F(s, \cos\Theta)$ holds, where the corresponding mean value $\bar{\Theta}(s)$ however may differ from the one obtained for $\operatorname{Im} F(s, \cos\Theta)$. Also $\bar{\Theta}(s)$ may change quite irregularly with varying s . Notice also that since the bound is proportional to ε^{-2} , it gets bigger if we attempt to reduce the interval III (9). III (10) implies the following:

If $|\operatorname{Im} F(s, \cos\theta)|$ is decreasing in $-2\varepsilon + \theta_0 < \theta < \theta_0 + 2\varepsilon$ then

$$\left| \frac{k}{s^{\frac{1}{2}}} \operatorname{Im} F(s, \cos(\theta_0 + 2\varepsilon)) \right| \leq \frac{1}{1 - \cos\varepsilon}$$

and one hopes that $\operatorname{Im} F(s, \cos\theta)$ is a slowly varying function of θ, i.e. there cannot appear great local oscillations. (This last property is typical for analytic, or even subharmonic functions.)

Consider now t in a fixed interval $t_1 \leq t \leq t_2$. For s sufficiently large we have

$$\theta_1 \cong \frac{|t_1|^{\frac{1}{2}}}{k} \quad ; \quad \theta_2 \cong \frac{|t_2|^{\frac{1}{2}}}{k} .$$

In order to apply III (10) we have to find θ and ε such that $\theta_1 = \theta - 2\varepsilon$ $\theta_2 = \theta + 2\varepsilon$. Thus $\theta = O(\frac{1}{s^{\frac{1}{2}}})$, $\varepsilon = O(\frac{1}{s^{\frac{1}{2}}})$ and there exists $\bar{t}(s)$ $(t_1 \leq \bar{t}(s) \leq t_2)$ such that

III (11) $$|\operatorname{Im} F(s, \bar{t}(s))| < \text{const} \cdot s .$$

We may compare this with the estimate obtained using analytic properties of $\operatorname{Im} F(s,t)$

III (12) $$|\operatorname{Im} F(s,t)| < \text{const} \cdot s (\log s)^2$$

for physical t .

We now turn to another property of the absorptive part $A_s(s, \cos\theta) = \operatorname{Im} F(s, \cos\theta)$ of the scattering amplitude. This property follows from unitarity and is another form of the positivity property III (3'). Let us consider the scattering in a plane. Let (θ_i, φ_i) be the angular variables of $\underset{\sim}{\Omega}_i$ $(i = 1, 2)$ ($\underset{\sim}{\Omega}_2$ being the angular direction of the initial wave packet and $\underset{\sim}{\Omega}_1$ the angular direction of the final wave packet in the center of mass system). We first remark that the left hand side of III (3) is positive. Choosing then for h in III (3) a function of the form

$$h(\underset{\sim}{\Omega}_i) = \delta_\varepsilon(\cos\theta_i)\, \tilde{g}(\varphi_i)$$

where δ_ε is an approximation of the δ - function and using

$$\underset{\sim}{\Omega}_1(\cos\theta_1 = 1, \varphi_1) \cdot \underset{\sim}{\Omega}_2(\cos\theta_2 = 1, \varphi_2) = \cos(\varphi_2 - \varphi_1)$$

we obtain in the limit $\varepsilon \to 0$:

$$\text{III(13)} \qquad \int \overline{\tilde{g}(\varphi_1)} \; \mathrm{Im}\, F(s, \cos(\varphi_2 - \varphi_1))\, \tilde{g}(\varphi_1)\, d\varphi_2\, d\varphi_1 \geq 0 \,.$$

We formulate our result as a

Theorem: $\mathrm{Im}\, F(s, \cos\Theta)$ is in Θ a function of positive type.

Bochner's theorem [Y 2] then implies that the Fourier coefficients $c_n'(s)$ in the expansion $\mathrm{Im}\, F(s, \cos\Theta) = \sum_{-\infty}^{\infty} c_n'(s)\, e^{in\Theta}$ satisfy $c_n'(s) \geq 0$. Indeed, choosing $\tilde{g}(\varphi) = e^{-in\varphi}$ in III (13) gives this statement immediately. Since furthermore $\cos\Theta$ is an even function in Θ, we have $c_{-n}'(s) = c_n'(s)$ such that finally

$$\text{III (14)} \qquad \mathrm{Im}\, F(s, \cos\Theta) = \sum_{n=0}^{\infty} c_n(s) \cos n\Theta \;; \quad c_n(s) \geq 0 .$$

Remark: Starting from

$$\mathrm{Im}\, F(s, \cos\Theta) = \frac{s^{\frac{1}{2}}}{2k} \sum_{l=0}^{\infty} (2l+1)\, \mathrm{Im}\, f_l(s)\, P_l(\cos\Theta)$$

and using [see Appendix]

$$P_l(\cos\Theta) = \sum_{n=0}^{\infty} c_{nl} \cos n\Theta \;; \quad c_{nl} \geq 0 ,$$

we immediately obtain III (14). In the above proof, however, we only used positivity and the rotational invariance of the scattering amplitude in a plane of the center of mass system. For the case of particles with spin, relations similar to III (14) may be proved if the initial and final states have the same helicity. An immediate consequence of III (14) is the

Corollary: For all $n \geq 0$ and $-1 \leq \cos\Theta \leq 1$

$$\text{III (15)} \qquad \left| \frac{d^n\, \mathrm{Im}\, F(s, \cos\Theta)}{d\cos\Theta^n} \right| \leq \frac{d^n}{d\cos\Theta^n}\, \mathrm{Im}\, F(s, \cos\Theta = 1)$$

For the proof we note that

$$\frac{d}{d\cos\Theta} \cos n\Theta = \frac{n \sin n\Theta}{\sin\Theta} = n\, \frac{e^{in\Theta} - e^{-in\Theta}}{e^{i\Theta} - e^{-i\Theta}} = 2n \left[\cos(n-1)\Theta + \cdots \right]$$

such that the derivative of an even function of positive type is again of positive type. Also we used the C^{∞} - property of $\operatorname{Im} F(s, \cos\Theta)$, which is guaranteed by the analyticity inside the large Lehmann ellipse [L 1] .

III b) Elastic unitarity. The construction of the amplitude

Before turning to the more complicated problem of the analyticity properties of the scattering amplitude we want to discuss another problem: The scattering of two scalar particles at an energy which is below the first inelastic threshold. Suppose we were able to measure with infinite accuracy the differential cross-section

$$\frac{d\sigma}{d\Omega} = \frac{1}{k^2} \left| \hat{F}(s, \cos\Theta) \right|^2 .$$

The question then is whether this fixes $\hat{F}(s, \cos\Theta)$ when unitarity is taken into account. It has been discussed by several authors. It is a problem which is generally thought to be trivial but which is in fact very difficult [C 4] [N 2] [M 12]. The problem we consider is an idealization of the problem of experimental phase shift analysis. It is, however, different because at the one hand we postulate infinite accuracy at a given energy but on the other hand we do not use Coulomb interference and continuity with respect to energy. The problem may be divided into two parts:

a) Existence ;

b) Uniqueness .

There is first the obvious ambiguity $\hat{F}(s, \cos\Theta) \rightarrow -\hat{F}^*(s, \cos\Theta)$ which amounts to replacing each phase shift by its negative in the partial wave expansion. A more subtle ambiguity has been found by Crichton [C 1] in the relatively simple case of a distribution with S- , P- and D- waves:

$$\delta_0 = -23^\circ 20' \; ; \quad \delta_1 = -43^\circ 27' \; ; \quad \delta_2 = 20^\circ$$

$$\delta_0' = 98^\circ 50' \; ; \quad \delta_1' = -26^\circ 33' \; ; \quad \delta_2 = 20^\circ$$

give the same angular distribution for the cross-section.

The amplitudes have the form

$$\frac{2}{15} e^{i\delta_2} \sin\delta_2 \left[\cos\Theta - z_1\right]\left[\cos\Theta - \frac{4}{5} + \frac{i \operatorname{ctg} \vartheta_2}{5}\right]$$

$$\frac{2}{15} e^{i\delta_2} \sin\delta_2 \left[\cos\Theta - \bar{z}_1\right]\left[\cos\Theta - \frac{4}{5} + \frac{i \operatorname{ctg} \vartheta_2}{5}\right],$$

where $\mathrm{Re}\, z_1$ and $|z_1|^2$ are given as functions of δ_2. Concerning the existence, we will find a sufficient condition, which is stronger than merely the requirement that the optical theorem should hold. On the other hand we will give an example, without solution, even though the optical inequality in the forward direction is satisfied.

In the following discussion, we will assume that $|\hat{F}(s,\cos\Theta)|$ is a bounded Hölder-continuous function of index $\frac{1}{2}$ in Θ :

$$\sup_{\Theta} |\hat{F}(s,\cos\Theta)| \leq C(s) < \infty,$$

$$\sup_{\Theta_1,\Theta_2} \frac{\left||\hat{F}(s,\cos\Theta_1)| - |\hat{F}(s,\cos\Theta)|\right|}{|\Theta_1-\Theta_2|^{\frac{1}{2}}} \leq C(s).$$

We shall write $\hat{F}(12)$ to denote $\hat{F}(s, \underset{\sim}{\Omega}_1\cdot\underset{\sim}{\Omega}_2)$ and

$$\hat{F}(12) = |\hat{F}(12)|\, e^{i\varphi(12)} .$$

Now we want to find $\varphi(12)$ such that

$$\text{III (16)} \quad \sin\varphi(12)\, |\hat{F}(12)| = \frac{1}{4\pi}\int d\Omega_3\, |\hat{F}(13)|\,|\hat{F}(32)| \cos(\varphi(13)-\varphi(32))$$

For given $|\hat{F}(12)|$ this is a nonlinear equation for $\varphi(12)$.
We shall show that the condition

$$\text{III (17)} \quad \underset{\underset{\sim}{\Omega}_1\cdot\underset{\sim}{\Omega}_2}{\mathrm{Max}} \frac{1}{4\pi}\frac{\int|\hat{F}(13)|\,|\hat{F}(32)|\, d\Omega_3}{|\hat{F}(12)|} = \sin\mu < 1$$

guarantees the existence of a solution. Equation III (17) gives $|\sin\varphi| < \sin\mu$ i.e. $-\mu < \varphi < \mu$ if we impose $\mathrm{Re}\,\hat{F}(s,\cos\Theta)\big|_{\cos\Theta=1} > 0$. Therefore sin φ uniquely determines φ in a continuous way. We may even prove that continuous solutions are necessarily such that under condition III (17) φ stays between 0 and μ. Indeed, assume $\varphi_{min} \leq \varphi \leq \varphi_{max}$ then if $(\varphi_{max}-\varphi_{min}) < \frac{\pi}{2}$ III (16) evidently gives sin $\varphi > 0$. On the other hand, if $(\varphi_{max}-\varphi_{min}) > \frac{\pi}{2}$ then cos $(\varphi_{max}-\varphi_{min}) < 0$ and III (16) gives

$$\sin\varphi(12) \geq \sin\mu\, \cos(\varphi_{max}-\varphi_{min}) .$$

In particular this holds for $\varphi = \varphi_{min}$ and we obtain

$$tg\, \varphi_{min} > \frac{\sin\mu \cdot \cos \varphi_{max}}{1 - \sin\mu \cdot \sin \varphi_{max}} > 0 ,$$

which is absurd, since $\varphi_{min} \leq 0$.

Since we want to discuss the nonlinear equation III (16), it is useful to consider the non-linear operator $\mathcal{O}$ defined by

$$\varphi' = \mathcal{O}(\varphi)$$

with

$$\text{III (18)} \qquad \sin \varphi'(12) = \frac{1}{4\pi} \frac{\int d\Omega_3 \, |\hat{\mathcal{F}}(13)| \, |\hat{\mathcal{F}}(32)| \cos(\varphi(13) - \varphi(32))}{|\hat{\mathcal{F}}(12)|}$$

defined for all continuous functions φ on [-1 , 1] with values in $[0, \mu]$. Then clearly $\mathcal{O}(\varphi)$ has the same properties as φ. Let $\mathcal{B}$ be the Banach space of all continuous real valued functions φ on [-1 , 1] with the norm

$$\| \varphi \| = \sup_{x \in [-1,1]} |\varphi(x)|$$

Clearly the set $\mathcal{C}_\mu$ consisting of all φ with values in $[0, \mu]$ is a bounded, closed convex set in $\mathcal{B}$ and $\mathcal{O}$ maps $\mathcal{C}_\mu$ into itself.
Let $\mathcal{C}_\mu$ have the topology induced by $\mathcal{B}$. We want to show that $\mathcal{O}$ is continuous. But this is trivial:

The estimate $|\cos a - \cos b| \leq |a - b|$ immediately gives

$$|\sin \mathcal{O}(\varphi)(12) - \sin \mathcal{O}(\psi)(12)| \leq 2 \sin\mu \, \| \varphi - \psi \| ; \; \varphi, \psi \in \mathcal{C}_\mu .$$

Since furthermore

$$|a - b| = \left| \int_a^b dx \right| = \left| \int_a^b \frac{dx}{d \sin x} \, d\sin x \right| \leq \frac{1}{\cos b} |\sin b - \sin a|$$

for $0 \leq a \leq b \leq \mu$, we finally have

$$\| \mathcal{O}(\varphi) - \mathcal{O}(\psi) \| \leq 2 \, tg\, \mu \, \| \varphi - \psi \| .$$

In particular for $\operatorname{tg}\mu < \frac{1}{2}$, $\mathcal{O}$ is a contraction and there exists a unique fixed point φ^* of $\mathcal{O}$ ([K 2] page 625) such that

$$\varphi^* = \mathcal{O}(\varphi^*) \quad ; \quad \varphi^* \in \mathcal{C}_\mu .$$

In the general case $\sin\mu < 1$ our intention is to apply the Leray-Schauder principle, which guarantees a fixed point for $\mathcal{O}$ ([K 2] page 645). For this we have to show that the convex closure of $\mathcal{O}^n(\mathcal{C}_\mu)$ is compact for at least one $n \geqslant 1$. The simplest way to show this is the use of the Ascoli-Arzela theorem. Now $\mathcal{O}(\mathcal{C}_\mu) \subset \mathcal{C}_\mu$ so the boundedness of $\mathcal{O}^n(\mathcal{C}_\mu)$ is trivial. Let us consider the equicontinuity: For a given

$\varphi \in \mathcal{C}_\mu$, set $\hat{F}_\varphi = |\hat{F}| e^{i\varphi}$ and

$$\hat{F}_\varphi(s, \cos\Theta) = \sum_{\ell=0}^{\infty} (2\ell+1) f_\ell^\varphi(s) P_\ell(\cos\Theta)$$

then

$$\operatorname{Im} \hat{F}_{\mathcal{O}(\varphi)}(s, \cos\Theta) = \sum_{\ell=0}^{\infty} (2\ell+1) |f_\ell^\varphi(s)|^2 P_\ell(\cos\Theta)$$

Now $f_\ell^\varphi(s)$ depends on φ but

$$\sum_{\ell=0}^{\infty} (2\ell+1) |f_\ell^\varphi(s)|^2 = \frac{1}{4\pi} \int d\Omega_2 \, |\hat{F}(12)|^2$$

depends only on $|\hat{F}|$. We use the inequality (see Appendix)

$$|P_\ell(\cos\Theta_1) - P_\ell(\cos\Theta_2)| \leq C \frac{\sqrt{|\Theta_1 - \Theta_2|}}{|\sin\Theta_1 \sin\Theta_2|^{\frac{1}{4}}}$$

which gives

$$|\operatorname{Im}\hat{F}_{\mathcal{O}(\varphi)}(s, \cos\Theta_1) - \operatorname{Im}\hat{F}_{\mathcal{O}(\varphi)}(s, \cos\Theta_2)| \leq \frac{C\sqrt{|\Theta_1 - \Theta_2|}}{|\sin\Theta_1 \sin\Theta_2|^{\frac{1}{4}}} \int \frac{d\Omega_3}{4\pi} |\hat{F}(13)|^2$$

Since the condition $\sin\mu < 1$ implies in particular that $|\hat{F}|$ never vanishes in $[-1, 1]$, the Hölder-continuity of $|\hat{F}|$ gives

III (19)

$$|\mathcal{O}(\varphi)(\cos\Theta_1) - \mathcal{O}(\varphi)(\cos\Theta_2)| \leq \frac{C\sqrt{|\Theta_1 - \Theta_2|}}{|\sin\Theta_1 \sin\Theta_2|^{\frac{1}{4}}}$$

where C does not depend on φ.
If we except $\Theta = 0$ and $\Theta = \pi$, III (19) is precisely an equicontinuity condition.
In order to get rid of the singularities at the extremities we iterate once and obtain

$$\mathrm{Im}\, \hat{F}_{O^2(\varphi)}(s, \cos\Theta_{12}) - \mathrm{Im}\, \hat{F}_{O^2(\varphi)}(s, \cos\Theta_{1'2})$$

$$= \frac{1}{4\pi} \int [\hat{F}_{O(\varphi)}(13) - \hat{F}_{O(\varphi)}(1'3)]\, \hat{F}^*_{O(\varphi)}(23)\, d\Omega_3$$

hence

$$|\mathrm{Im}\, \hat{F}_{O^2(\varphi)}(s, \cos\Theta_{12}) - \mathrm{Im}\, \hat{F}_{O^2(\varphi)}(s, \cos\Theta_{1'2})|$$

$$\leq C \int \frac{\sqrt{|\Theta_{13} - \Theta_{1'3}|}}{|\sin\Theta_{13} \cdot \sin\Theta_{1'3}|^{\frac{1}{4}}} |\hat{F}(32)|\, d\Omega_3 \leq C\sqrt{|\Theta_{11'}|} \int \frac{|\hat{F}(32)|\, d\Omega_3}{|\sin\Theta_{13} \cdot \sin\Theta_{1'3}|^{\frac{1}{4}}}$$

$$\leq C\sqrt{|\Theta_{11'}|},$$

since $|\hat{F}(32)|$ is bounded. This proves that $O^2(\mathcal{C}_\mu)$ consists of equicontinuous functions and is therefore precompact. The convex closure of $O^2(\mathcal{C}_\mu)$ which is contained in $\mathcal{C}_\mu$ is then compact and we may apply the Leray-Schauder theorem. Thus we have established the existence of a fixed point for $\sin\mu < 1$.
Let us now consider uniqueness. For $\mathrm{tg}\,\mu < \frac{1}{2}$ we used the fact that O was a contraction. We will now prove that for

$$2\, \mathrm{tg}\,\mu \cdot \sin\mu < 1$$

O still is a contraction. Indeed let us consider the Fréchet derivative O'_φ of O at the point φ. O'_φ is a linear operator, whose action on $\psi \in \mathcal{B}$ is described by

$$O'_\varphi(\psi)(12) = -\frac{\int d\Omega_3\, |\hat{F}(13)||\hat{F}(32)| \sin(\varphi(13) - \varphi(32))\, (\psi(13) - \psi(32))}{\cos O(\varphi)(12) \cdot 4\pi \cdot |\hat{F}(12)|}.$$

Thus we get for the operator norm

$$\| O'_\varphi \| < 2\, \mathrm{tg}\,\mu\, \sin\mu \quad ; \quad \varphi \in \mathcal{C}_\mu$$

since $|\sin(\varphi(13)-\varphi(32))| < \sin\mu$ for $\varphi \in \mathcal{C}_\mu$. But $\mathcal{O}$ is a contraction on $\mathcal{C}_\mu$ if $\| \mathcal{O}'_\varphi \| < 1$, i.e. if $2\,\mathrm{tg}\,\mu \cdot \sin\mu < 1$, which is what we wanted to prove ([K2] page 661). By a more subtle discussion, it is possible to improve the uniqueness bound to $\mathrm{tg}\,\mu \cdot \sin\mu < 1$ which corresponds to $\sin\mu < 0.79$. The examination of the case of a finite, but arbitrary large number of partial waves makes it, however, very likely that the solution is still unique for $0.79 \le \sin\mu < 1$. Though the condition

$$\text{III (20)} \qquad \sup_{\underset{\sim}{\Omega}_1, \underset{\sim}{\Omega}_2} \frac{1}{4\pi} \frac{\int |\hat{F}(13)|\,|\hat{F}(32)|\, d\Omega_3}{|\hat{F}(12)|} < 1$$

was only shown to be a sufficient condition for the existence of a solution, it is clear that the numerical constant, 1, cannot be changed because in forward direction the condition

$$\text{III (21)} \qquad \frac{1}{4\pi} \frac{\int |\hat{F}(13)|^2 \, d\Omega_3}{|\hat{F}(11)|} < 1$$

is necessary. On the other hand, condition III (20) is physically not satisfactory since this imposes $\mathrm{Im}\, f_\ell < \frac{1}{2}$ for $\ell > 0$, so there may appear only S-wave resonances [M 12] .

Let us now give an example where III (21) is satisfied, but for which there is no solution.

Take $|\hat{F}(s, \cos\Theta)| = \lambda f(\Theta)$ where $f(\pi) = 0$ and $\frac{d}{d\cos\Theta} f(\Theta) > 0$ for all Θ. Then for λ small enough there is no acceptable unitary amplitude: We have

$$|\sin\varphi(12)| < \frac{\lambda}{4\pi} \frac{\int f(23)\, f(31)\, d\Omega_3}{f(12)} .$$

For $\cos\Theta_{12} > -1 + \lambda^{\frac{1}{2}}$ we have $f(12) > c\,\lambda^{\frac{1}{2}}$, hence $|\sin\varphi(12)| < c\,\lambda^{\frac{1}{2}}$ or $|\varphi(12)| < c\,\lambda^{\frac{1}{2}}$. This gives

$$\mathrm{Im}\, \hat{F}(s, \cos\Theta = -1) \ge \frac{\lambda}{2} \int_{-1+\lambda^{\frac{1}{2}}}^{1-\lambda^{\frac{1}{2}}} f(\Theta)\, f(-\Theta+\pi)\,(1 - c'\lambda)\, d\cos\Theta$$

$$- \lambda \int_{1-\lambda^{\frac{1}{2}}}^{1} f(\Theta)\, f(-\Theta+\pi)\, d\cos\Theta .$$

For λ small enough, the right hand side is positive, hence $|\Phi(s, \cos\Theta = -1)|$ cannot vanish, which is a contradiction.

IV. Review of the results of local field theory on analytic properties of scattering amplitudes on the mass shell

Dispersion relations were first used in physics by Kramers and Kronig who connected the absorption of light and the index of refraction of light in matter through dispersion relations. However, it was only in 1955 that Goldberger wrote a dispersion relation for the pion nucleon scattering [G 3]. Later on these dispersion relations were proved for the forward scattering amplitude by K. Symanzik [S 9] and in the general case by H.J. Bremermann, R. Oehme, J.G. Taylor [B 9] on the one hand and N.N. Bogoliubov, B.V. Medvedev, M.K. Polivanov [B 6] on the other hand from the principles of local field theory as stated by Lehmann, Symanzik and Zimmermann. In a few years considerable progress was made. However, this golden period of axiomatic field theory ended in 1958 with the work of Lehmann [L 1]. Since then progress has been very slow.

The first, by now classical, result is the dispersion relation for fixed t : The scattering amplitude is for $t_0 \leq t \leq 0$ the boundary value of real analytic function

$$F(s,t) = F^*(s^*,t)$$

holomorphic in the s-plane except for poles related to stable particles and cuts along real s :

$$s \geq s_{thr} \quad ; \quad s \leq -t + 2M_A^2 + 2M_B^2 - u_{thr}$$

such that

$$\lim_{\varepsilon \to 0^+} F(s+i\varepsilon, t) = F_{AB \to AB}(s,t) ; \; s \geq (M_A + M_B)^2.$$

(In future we will drop the pole term, since it gives only an unessential complication.) The left hand cut is connected with the process in the crossed channel, which reflects the usefulness of the variable u . Along the left hand cut

$$\lim_{\varepsilon \to 0^+} F(s-i\varepsilon, t) = F_{A\bar{B} \to A\bar{B}}(u,t) ; \; u \geq (M_A + M_B)^2$$

where the square of the center of mass energy u of the reaction $A + \bar{B} \to A + \bar{B}$ is given by $u = 2M_A^2 + 2M_B^2 - s - t$. The square of the momentum transfer is again t . In addition, the discontinuities of F across the cuts are given by the absorptive parts in the s- and u-channel respectively. These analyticity properties hold for a number of processes such as

$$\pi N \to \pi N \ , \ \pi\pi \to \pi\pi \ , \ \pi\pi \to K\bar{K} \ , \ \pi K \to \pi K$$

$$KK \to KK \ , \ \pi\Lambda \to \pi\Lambda \ , \ \pi\Sigma \to \pi\Sigma$$

Of these reactions, only $\pi N \to \pi N$ is physically observable. On the other hand, processes such as $NN \to NN$ or $KN \to KN$ are not on the list. There is very little hope for $NN \to NN$ while for $KN \to KN$ something might be done.

These fixed t analyticity properties are not quite enough to write a dispersion relation, i.e. a Cauchy integral. What we need in addition is that the scattering amplitude is polynomially bounded. This is true in the L.S.Z formalism and also starting from the Wightman axioms [H 1] . In the theory of local observables as formulated by Araki as well as in Jaffe's theory, where the fields also are local but not necessarily give tempered distributions, this has only recently been proved [E 2].

Assuming the polynomial boundedness, we may write

$$\text{IV (1)} \qquad F(s,t,u) = \frac{s^N}{\pi} \int_{s_{thr}}^{\infty} \frac{A_s(s',t)}{s'^N(s'-s)}\, ds' + \frac{u^N}{\pi} \int_{u_{thr}} \frac{A_u(u',t)}{u'^N(u'-u)}\, du'$$

$$+ \text{ Polynomial in } s \text{ and } u$$

for a suitable N .

Now how can we interpret the boundary values of $F(s,t,u)$ on the cuts? We said that it was the physical scattering amplitude. This holds, however, only in the good cases where we have the normal thresholds $s_{thr} = u_{thr} = (M_A + M_B)^2$. However, for $t < 0$ there is a difficulty because from the relation $t = -2k^2(1-\cos\Theta)$, we see that if $\cos\Theta$ is to stay in the physical region, then as $s \to s_{thr}$, $k^2 \to 0$ and hence $t \to 0$. So if we hold t fixed and negative, $\cos\Theta$ can become much less than -1 as $s \to s_{thr}$.

Fortunately Lehmann [L 1] was able to give a meaning to $F(s,t,u)$ for s physical and $\cos\Theta$ outside the interval $-1 \le \cos\Theta \le 1$. He proved that $F(s,\cos\Theta)$ is analytic inside an ellipse in the complex $\cos\Theta$ - plane with foci at ± 1 and semimajor axis

$$\text{IV (2)} \qquad \cos\Theta_0(s) = \left[1 + \frac{(M_A'^2 - M_A^2)(M_B'^2 - M_B^2)}{k^2(s-(M_A'-M_B')^2)} \right]^{\frac{1}{2}}$$

where M'_A is the state of lowest mass such that $\langle A' | j_A(0) | 0 \rangle \neq 0$ and similarly for M'_B ; e.g. G-parity considerations give $|A'\rangle = |3\pi\rangle$ for $j_\pi(0)$ and for $j_N(0)$ we have $|A'\rangle = |N\pi\rangle$. This is still not quite enough because for $k \to 0$ it only allows the continuation of $\mathcal{F}(s,t)$ to $t = -2k^2(1-\cos\theta_0) \cong k^2 \frac{\text{const}}{k}$ which again goes to zero with k.

However, Lehmann also proved that the absorptive part $A_s(s, \cos\theta)$ is analytic in a bigger confocal ellipse with semi-major axis $2\cos^2\theta_0(s) - 1$. Then $A_s(s,\cos\theta)$ is still defined for $\cos\theta > -(2\cos^2\theta_0(s)-1)$ i.e. for t in

$$0 \geq t \geq -4k^2\cos^2\theta_0(s) = -\left[4k^2 + 4\,\frac{(M_A'^2 - M_A^2)(M_B'^2 - M_B^2)}{s - (M'_A - M'_B)^2}\right].$$

Taking the maximum of the left hand side gives: $A_s(s,t)$ is defined for all physical s and all t in the interval

$$0 \geq t \geq t_0$$

where $t_0 = -28\mu^2$ for pion-pion scattering and $t_0 = -12.4\mu^2$ for pion-nucleon scattering (μ = pion mass). For these t the integral IV (1) is well defined and hence also $\mathcal{F}(s,t)$. These results are still too weak for the following reasons:

(i) $\mathcal{F}(s,t)$ has analytic properties in s for t real < 0 (the fixed dispersion relation) and analytic properties in t (the Lehmann ellipse) but we know nothing yet about the analyticity properties of $\mathcal{F}(s,t)$ for s and t both complex.

(ii) In cases where we have no dispersion relations as for the NN $\longrightarrow$ NN scattering, nothing survives, and in particular it is no longer clear whether or not the particle-particle scattering amplitude can be continued to the particle-antiparticle amplitude (crossing property), a property which is essential for proving, for instance, the Pomeranchuk theorem.

(iii) The Lehmann ellipse is too small. To see this, we remark that [see Appendix]

$$\sum_{\ell=0}^{\infty} a_\ell P_\ell(z)$$

is absolutely convergent in an ellipse with foci ± 1 and semimajor axis $z_0 > 1$ if and only if

$$\text{IV (3)} \qquad \lim_{\ell} |a_\ell|^{\frac{1}{\ell}} \leq \frac{1}{z_0 + \sqrt{z_0^2 - 1}}.$$

Now for the absorptive part we have $z_0 \cong 1 + \frac{\text{const}}{s^2}$ and hence

$$\overline{\lim_{\ell}} \, |\text{Im}\, f_\ell(s)|^{\frac{1}{\ell}} \leq \frac{1}{1 + \frac{\text{const}}{s}} \cong e^{-\frac{\text{const}}{s}} .$$

Thus if $\text{Im}\, f_\ell(s)$ is sufficiently smooth, we have $\text{Im}\, f_\ell(s) \sim e^{-\text{const}\frac{\ell}{s}}$ times a smooth function in s and ℓ.

This means that the number of partial waves which contribute to the forward scattering is of the order of $L_{max} \sim s$. This is not acceptable because

(a) it gives bounds for the forward scattering amplitude that are too high (Greenberg-Low),

(b) it does not fit with the general idea that the ranges of the forces between elementary particles are energy independent and are given by the Compton wave-length of the lightest object which these particles can exchange. By a well known semiclassical argument [S 1], we know that if the range of the forces is R then $L_{max} \cong k \cdot R$ so that

$$L_{max} \cong \text{const} \sqrt{s}$$

Let us discuss the remedies to (i) and (iii). Mandelstam [M 3] was the first to prove analyticity in both s and t. His argument applied to the $\pi - \pi$ amplitude for which dispersion relations exist in all three channels, but it was later generalized by Lehmann [L 2] to situations where only a fixed t dispersion relation is given as e.g. for the pion-nucleon scattering. The analyticity domain for the $\pi - \pi$ case is given by $|st| < 256\mu^4$. Inside this analyticity domain the only singularities are the physical cuts on the real axes: s, t or $u \geq 4\mu^2$. This domain unfortunately does not contain the whole physical region. A big step forward was then made by Bros, Epstein and Glaser [B 10] who showed that given any two body reaction $A + B \longrightarrow C + D$ where A, B, C, D are stable particles, then any point $s_0, \cos\theta_0$ in the physical region is surrounded by an analyticity neighborhood

IV (4)
$$|s - s_0| < \eta(s_0, \cos\theta_0)$$
$$|\cos\theta - \cos\theta_0| < \eta(s_0, \cos\theta_0)$$

the only singularities being given by the cut $(M_A + M_B)^2 \leq s \leq \infty$. In contrast to the work of Mandelstam and Lehmann, in which the on-shell amplitude was discussed, the latter result was obtained starting from the initial analyticity domain of the off-shell amplitude. The only weakness of the domain in III (4) is that η as a function of s_0 and $\cos\theta_0$ is not explicitly known.

The answer to point (ii) was also given by Bros, Epstein and Glaser [B 11], who proved that it is always possible to continue analytically the mass-shell particle-particle amplitude to the particle-antiparticle amplitude. More precisely they show that given any $t \leq 0$ the scattering amplitude is analytic in the cut s-plane except for possible singularities in a finite region. The extension of this region goes as $|t|^{3+\varepsilon}$ [E3]. An estimate of the form $|t|$ would, however, be welcome, since this would enlarge the analyticity domain of the partial wave amplitudes.

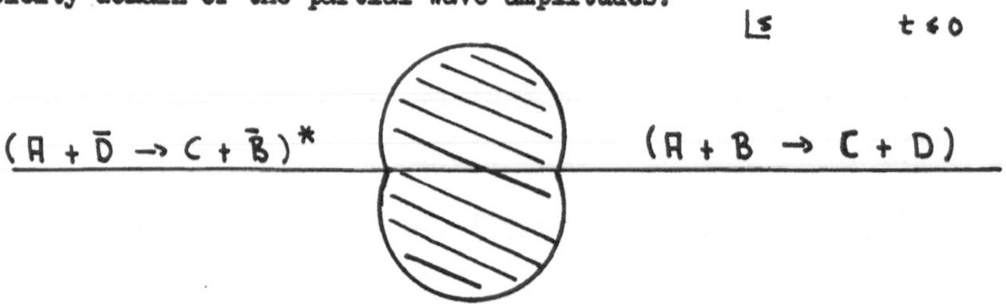

It is clear that by avoiding these singularities we may continue from the particle-particle amplitude to the complex conjugate of the particle-antiparticle amplitude.
Using the path shown below,

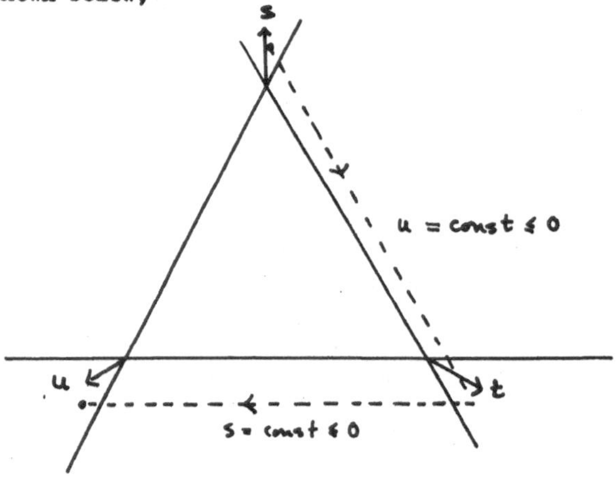

we also arrive at a point below the left hand cut in the s-plane, since along this path the scattering amplitude is complex-conjugated twice.

These results on the analyticity domain are sufficient to prove the Pomeranchuk theorem. That is to say the proof of the Pomeranchuk theorem is as good (or as bad) for NN as for πN scattering even though we do not have a dispersion relation for NN scattering. More precisely:

If $$\lim_{s \to \infty} \left[\sigma_{tot}(AB) - \sigma_{tot}(A\bar{B}) \right]$$

exists, the limit being finite or infinite, and if the forward amplitude is such that

$$\lim_{s \to \infty} \frac{F(s, t=0)}{s \log s} = 0,$$

then actually the above limit is equal to zero [M 7].
As to (iii), we shall see in the next section that we indeed can enlarge the Lehmann ellipse by using the positivity property III (15) of $A_s(s,t)$.

V. Extension of the axiomatic analyticity domain of scattering amplitudes by unitarity

In this section we shall use the positivity property to improve the results quoted in the preceeding section. We consider the elastic scattering amplitude of spinless particles not necessarily having the same mass. The case with spin will be considered in section XIV. Also let us assume fixed t dispersion relations. Later, however, we will show how to dispense with this condition.
Our intention will be to prove combined analyticity in s and in t. Now if we are able to expand the amplitude into the following series

$$F(s,t) = \sum_{n=0}^{\infty} \frac{t^n}{n!} \left(\frac{d}{dt}\right)^n F(s,0),$$

then according to Hartog's theorem (see e.g. [H 2]), $F(s,t)$ will be analytic in a domain $\{(s,t) \mid s \in \mathcal{D}, |t| < R\}$ if all $\left(\frac{d}{dt}\right)^n F(s,0)$ are analytic in $\mathcal{D}$ and if the above series is absolutely convergent for $|t| < R$ uniformly for s in an arbitrary compact subset of $\mathcal{D}$. We will therefore try to find a bound for $\left(\frac{d}{dt}\right)^n F(s,0)$.

In order to exhibit the arguments more clearly, let us start with an unsubtracted dispersion relation without a left hand cut. Then for a fixed t $(t_0 \leq t \leq 0)$ we have

V (1)
$$F(s,t) = \frac{1}{\pi} \int_{(M_A+M_B)^2}^{\infty} \frac{A_s(s',t)}{(s'-s)} ds'.$$

Now the work of Bros, Epstein and Glaser [B 10] or of Lehmann [42] guarantees the existence of a real point s_0 below, but near threshold such that $F(s,t)$ is analytic in (s,t) in a neighborhood of $s = s_0$, $t = 0$ (see IV (4)).
The absorptive part $A_s(s',t)$ is a distribution in s'. Let $\omega(x)$ be a positive C^∞ - function with support in $-x_0 \leq x \leq x_0$ where $0 < 2x_0 < (M_A+M_B)^2 - s_0$.
Put

$$F^\omega(s,t) = \int_{-x_0}^{+x_0} \omega(x) F(s-x,t)\, dx$$

V (2)
$$A_s^\omega(s',t) = \int_{-x_0}^{+x_0} \omega(x) A_s(s'-x,t)\, dx.$$

Then

(α) we still have a dispersion relation for t in $t_0 \leq t \leq 0$

V (3) $$\Phi^{\omega}(s,t) = \frac{1}{\pi} \int_{(M_A+M_B)^2 - x_0}^{\infty} \frac{A_s^{\omega}(s',t)}{s'-s} \, ds'.$$

(β) $\Phi^{\omega}(s,t)$ is analytic in (s,t) in a neighborhood

$|s - s_0| < \varepsilon(s_0)$, $|t| < R(s_0)$ of $s = s_0, t = 0$

for suitable $\varepsilon(s_0)$ and $R(s_0)$.

(γ) $A_s^{\omega}(s',t)$ is C^{∞} in s' and t for

$$(M_A + M_B)^2 - x_0 \leq s < \infty, \quad t_0 \leq t \leq 0$$

because of the existence of a large Lehmann-ellipse.

It is also not hard to see that the positivity property still holds:

V (4) $$\left(\frac{d}{dt}\right)^n A_s^{\omega}(s',0) \geq \left|\left(\frac{d}{dt}\right)^n A_s^{\omega}(s',t)\right|_{-4k'^2 + x_0 \leq t \leq 0}.$$

We now want to take the t-derivative in V(3) at the point $s = s_0$, $t = 0$. The question that immediately arises is whether we may interchange differentiation and integration. Here is where the positivity property V (4) essentially comes in. Since the derivative exists, we are free to define it as

V (5) $$\frac{d}{dt}\Phi^{\omega}(s_0,0) = \lim_{\tau \to 0^+} \frac{\Phi^{\omega}(s_0,0) - \Phi^{\omega}(s_0,-\tau)}{\tau}$$

$$= \lim_{\tau \to 0^+} \frac{1}{\pi} \int_{(M_A+M_B)^2 - x_0}^{\infty} \frac{A_s^{\omega}(s',0) - A_s^{\omega}(s',-\tau)}{\tau(s' - s_0)} \, ds'.$$

Note that $A_s^{\omega}(s',-\tau)$ is defined for sufficiently small τ because of the existence of the large Lehmann ellipse. We divide the integral into two parts:

$$\int_{(M_A+M_B)^2-x_0}^{\infty} = \int_{(M_A+M_B)^2-x_0}^{X} + \int_{X}^{\infty} ,$$

$$X > (M_A + M_B)^2 + x_0 .$$

Also Rolle's theorem gives

V (6) $$\frac{A_s^\omega(s',0) - A_s^\omega(s',-\tau)}{\tau} = \frac{d}{dt} A_s^\omega(s', -\tau(s'))$$

$$0 \leq \tau(s') \leq \tau .$$

Therefore, because of the regularization, the integrand in the first integral may be uniformly bounded, and since the integrand also converges for $\tau \to 0$, we may apply Lebesgues' uniform boundedness theorem. This gives

V (7) $$\lim_{\tau \to 0^+} \frac{1}{\pi} \int_{(M_A+M_B)^2-x_0}^{X} \frac{A_s^\omega(s',0) - A_s^\omega(s',-\tau)}{\tau(s'-s_0)} ds' = \frac{1}{\pi} \int_{(M_A+M_B)^2-x_0}^{X} \frac{\frac{d}{dt} A_s^\omega(s',0)}{s'-s_0} ds' .$$

On the other hand, the integrand in the second term $\int_X^\infty$ is positive for sufficiently small τ. To see this we note that since $X > (M_A+M_B)^2 + x_0$, for τ in the interval $-k^2(X) + x_0 \leq -\tau \leq 0$ we may apply V (4) for $n = 0$

$$A_s^\omega(s',0) - A_s^\omega(s',0) \geq 0 \; ; \quad X \leq s' < \infty$$

Thus, since $s' - s_0 > 0$ for all s'

$$\frac{d}{dt} F^\omega(s_0, 0) \geq \frac{1}{\pi} \int_{(M_A+M_B)^2-x_0}^{X} \frac{\frac{d}{dt} A_s^\omega(s',0)}{s'-s_0} ds' .$$

We let X tend to infinity:

V (8)
$$\frac{d}{dt} \Psi^{\omega}(s_0, 0) \geq \frac{1}{\pi} \int_{(M_A+M_B)^2 - x_0}^{\infty} \frac{\frac{d}{dt} A_s^{\omega}(s', 0)}{s' - s_0} \, ds'.$$

Note that the right hand side exists since the integrand is positive. To obtain equality we only have to prove the opposite inequality: Choosing X in the above decomposition fixed, we may again apply the positivity property V (4) for τ sufficiently small, this time for $n = 1$, which gives

V (9)
$$\frac{1}{\pi} \int_X^{\infty} \frac{A_s^{\omega}(s', -\tau(s'))}{s' - s_0} \, ds' \leq \frac{1}{\pi} \int_X^{\infty} \frac{\frac{d}{dt} A_s^{\omega}(s', 0)}{s' - s_0} \, ds'.$$

V (9) combined with V (7) gives the opposite inequality to V (8) so

V (10)
$$\frac{d}{dt} \Psi^{\omega}(s_0, 0) = \frac{1}{\pi} \int_{(M_A+M_B)^2 - x_0}^{\infty} \frac{\frac{d}{dt} A_s^{\omega}(s', 0)}{s' - s_0} \, ds'.$$

In order to iterate this procedure, we consider t in the interval $Max(-R(s_0), t_0) < t \leq 0$. Then

$$\frac{d}{dt} \Psi^{\omega}(s_0, t) = \lim_{\tau \to 0^+} \frac{\Psi^{\omega}(s_0, t) - \Psi^{\omega}(s_0, t-\tau)}{\tau}$$
$$= \lim_{\tau \to 0^+} \frac{1}{\pi} \left(\int_{(M_A+M_B)^2 - x_0}^{X} + \int_X^{\infty} \right) \frac{\frac{d}{dt} A_s^{\omega}(s', \tau(s'))}{(s' - s_0)} \, ds'$$
$$t - \tau \leq \tau(s') \leq t.$$

The first integral exists because of (γ). The second term may be estimated by

$$\left| \int_X^{\infty} \frac{\frac{d}{dt} A_s^{\omega}(s', \tau(s'))}{s' - s_0} \, ds' \right| \leq \int_X^{\infty} \frac{\left| \frac{d}{dt} A_s^{\omega}(s', \tau(s')) \right|}{s' - s_0} \, ds' \leq \int_X^{\infty} \frac{\frac{d}{dt} A_s^{\omega}(s', 0)}{s' - s_0} \, ds'$$

for $-4k^2(X) + x_0 \leq Max(-R(s_0), t_0) < t - \tau.$

This term therefore goes to zero as $X \to \infty$ since V (10) is finite.
The result is

$$\frac{d}{dt} F^{\omega}(s_0, t) = \frac{1}{\pi} \int_{(M_A+M_B)^2 - x_0}^{\infty} \frac{\frac{d}{dt} A_s^{\omega}(s', t)}{s' - s_0} ds',$$

which upon iteration gives

V (11) $$\left(\frac{d}{dt}\right)^n F^{\omega}(s_0, t) = \frac{1}{\pi} \int_{(M_A+M_B)^2 - x_0}^{\infty} \frac{\left(\frac{d}{dt}\right)^n A_s^{\omega}(s', t)}{s' - s_0} ds',$$

$$\mathrm{Max}\,(-R(s_0), t_0) < t \leq 0 .$$

This permits the following extension to complex s in the cut-plane and for $t = 0$:

$$\left(\frac{d}{dt}\right)^n F^{\omega}(s, 0) \underset{\mathrm{def.}}{=} \frac{1}{\pi} \int_{(M_A+M_B)^2 - x_0}^{\infty} \frac{\left(\frac{d}{dt}\right)^n A_s^{\omega}(s', 0)}{s' - s} ds',$$

where the positivity gives the estimate

$$\left| \left(\frac{d}{dt}\right)^n F^{\omega}(s, 0) \right| \leq \mu(s_0, s) \left(\frac{d}{dt}\right)^n F^{\omega}(s_0, 0)$$

with

$$\mu(s_0, s) = \underset{(M_A+M_B)^2 - x_0 \leq s' < \infty}{\mathrm{Max}} \frac{|s' - s_0|}{|s' - s|} \leq \frac{|s - s_0|}{|\mathrm{Im}\, s|}$$

For s not on the cut, one has $\mu(s_0, s) < \infty$.

Therefore the series

$$F^{\omega}(s, t) \underset{\mathrm{def.}}{=} \sum_{n=0}^{\infty} \frac{t^n}{n!} \left(\frac{d}{dt}\right)^n F^{\omega}(s, 0)$$

satisfies the estimate

$$\text{V (12)} \qquad |F^{\omega}(s,t)| \leq \mu(s_0,s) \sum_{n=0} \frac{|t|^n}{n!} \left(\frac{d}{dt}\right)^n F^{\omega}(s_0,0)$$

$$= \mu(s_0,s)\, F^{\omega}(s_0,|t|)$$

The conditions for applying Hartog's theorem are thus fulfilled, so we have proved that $F^{\omega}(s,t)$ is an analytic function in $(s,t) \in \mathcal{B}_{x_0}(R(s_0))$

$$\mathcal{B}_{x_0}(R(s_0)) = \{(s,t) \mid s \notin [(M_A + M_B)^2 - x_0, \infty),\ |t| < R(s_0)\}.$$

By the identity theorem for analytic functions this is indeed an extension of the original $F^{\omega}(s,t)$. To obtain similar statements for the unregularized amplitude, we let ω tend to the δ - function. Then $F^{\omega}(s_0,|t|) \to F(s_0,|t|)$ again because of (β). But then V(12) shows that F^{ω} is a normal family of analytic functions. We may therefore choose a subsequence which converges uniformly on any compact subset of $\mathcal{B}_0(R(s_0))$ to an analytic function. This limit function coincides with $F(s,t)$ in a neighborhood of $s = s_0$, $t = 0$, which shows uniqueness, $\mathcal{B}_0(R(s_0))$ being simply connected. Thus as a final result we have obtained an analytic continuation of $F(s,t)$ to $\mathcal{B}_0(R(s_0))$ with the estimate

$$\text{V (13)} \qquad |F(s,t)| \leq \mu(s_0,s)\, F(s_0,|t|),$$

or from the estimate of μ : $\quad |F(s,t)| < C(|t|) \cdot \frac{|s - s_0|}{|\operatorname{Im} s|}$.
Therefore the boundary value is a distribution in s for all t $(|t| < R(s_0))$ [S 8]. This indicates that we might extend the dispersion relation V (1) to all t $(|t| < R(s_0))$. Before doing this, we will from now on assume that the absorptive parts $A_s(s',t)$ and $A_u(u',t)$ themselves have the property (γ). This assumption is inessential since we saw above how to deal with the regularization, but it will exempt us from carrying the regularization through all arguments.

Now consider the point (s_0, t) $(0 \leq t < R(s_0))$. Then

$$F(s_0,t) = \sum_{n=0}^{\infty} \frac{t^n}{n!} \left(\frac{d}{dt}\right)^n F(s_0,0) \quad =$$

$$= \sum_{n=0}^{\infty} \frac{t^n}{n!} \int_{(M_A+M_B)^2}^{\infty} \frac{\left(\frac{d}{dt}\right)^n A_s(s',0)}{s'-s_0} \, ds' .$$

Since the integrand is positive, we may interchange summation and integration due to the theorem of Fubini-Tonelli [Y 1] . Thus

V (14) $$A_s(s',t) \underset{def}{=} \sum_{n=0}^{\infty} \frac{t^n}{n!} \left(\frac{d}{dt}\right)^n A_s(s',0)$$

is for fixed t, $0 \leq t < R(s_0)$ a positive measurable function in s' and we have

V (15) $$F(s_0,t) = \frac{1}{\pi} \int_{(M_A+M_B)^2}^{\infty} \frac{A_s(s',t)}{s'-s_0} \, ds'.$$

Now we may extend the definition V (14) to all t $(|t| < R(s_0))$, because positivity gives

V (16) $$|A_s(s',t)| \leq A_s(s',|t|).$$

This again permits us to extend V (15) to all t $(|t| < R(s_0))$. The final extension to s in the cut plane is then also possible:

V (17) $$F(s,t) = \frac{1}{\pi} \int_{(M_A+M_B)^2}^{\infty} \frac{A_s(s',t)}{s'-s} \, ds',$$

since V (16) again leads to the estimate V (13) for the right hand side and because we may apply the identity theorem for analytic functions. Note that s_0 in V (15) may vary in a small interval. In this way we have extended the validity of the dispersion relation V (1) to $\mathcal{G}_0(R(s_0))$ <u>without introducing a subtraction</u>.

Now let us consider the case where we have to start with a dispersion relation including subtractions. Let s_1 $(< s_0)$ be sufficiently close to s_0 . Consider for $t_0 \leq t \leq 0$

V (18) $$\phi(s,t;s_1) = F(s,t) - \sum_{n=0}^{N-1} \frac{(s-s_1)^n}{n!} \left(\frac{d}{ds}\right)^n F(s_1,t) =$$

$$= \frac{(s-s_1)^N}{\pi} \int_{(M_A+M_B)^2}^{\infty} \frac{A_s(s',t)}{(s'-s_1)^N (s'-s)} ds'.$$

Then the discussion above may immediately be transcribed to $\Phi(s,t;s_1)$. $\Phi(s,t;s_1)$ may be extended to an analytic function in (s,t) in $\mathcal{E}_0(R(s_0))$. Since $\left(\frac{d}{ds}\right)^n \mathcal{F}(s_1,t)$ is analytic in t $(|t| < R(s_0))$, $\mathcal{F}(s,t)$ also has an analytic continuation to $\mathcal{E}_0(R(s_0))$. The validity of the dispersion relation V (18) may also be extended to all points $(s,t) \in \mathcal{E}_0(R(s_0))$, $A_s(s',t)$ being defined by V (14).

Let us now turn to the more realistic case where we also have a left hand cut. We start with the case where a dispersion relation may be written without subtractions:

$$\text{V (19)} \quad \mathcal{F}(s,t) = \frac{1}{\pi} \int_{(M_A+M_B)^2}^{\infty} \frac{A_s(s',t)}{(s'-s)} ds' + \frac{1}{\pi} \int_{(M_A+M_B)^2}^{\infty} \frac{A_u(u',t)}{(u'-u)} du';$$

$$t_0 \leq t \leq 0 \;;\; s+u+t = 2M_A^2 + 2M_B^2.$$

We write the positivity conditions as

$$\left(\frac{d}{dt}\right)^n A_s(s',0) \geq \left| \left(\frac{d}{dt}\right)^n A_s(s',t) \right|_{\substack{-\varepsilon' \leq t \leq 0 \\ s' > (M_A+M_B)^2 + \varepsilon}} ,$$

$$\left(\frac{d}{dt}\right)^n A_u(u',0) \geq \left| \left(\frac{d}{dt}\right)^n A_u(u',t) \right|_{\substack{-\varepsilon' \leq t \leq 0 \\ u' > (M_A+M_B)^2 + \varepsilon}}$$

In order to apply these inequalities we choose s_0 real and between the two cuts:

$$(M_A - M_B)^2 < s_0 < (M_A + M_B)^2.$$

Since, however, $\mathcal{F}(s_0,t)$ has a singularity at $t = (M_A - M_B)^2 - s_0$ it turns out that it will be useful to discuss

$$\phi(s_0,t) = \frac{F(s_0,t)}{s_0-(M_A-M_B)^2-t-2\eta},$$

where η is chosen such that $s_0 > (M_A-M_B)^2+2\eta$ and $\varepsilon' > \eta > 0$. Before we start with the discussion of $\phi(s_0,t)$ let us make a remark: Let f be a function defined on $[-\eta, 0]$. f is said to have the property (P) if

$$\left|\left(\frac{d}{dt}\right)^n f(t)\right|_{-\eta\le t\le 0} \le \left(\frac{d}{dt}\right)^n f(0).$$

It is easy to see that fg has the property (P) if f and g have the property (P).

From inspection of

$$\text{V (20)} \quad \phi(s_0,t) = \frac{1}{\pi}\int_{(M_A+M_B)^2}^{\infty} \frac{A_s(s',t)\,ds'}{(s'-s_0)(s_0-(M_A-M_B)^2-\eta-(t+\eta))}$$

$$+\frac{1}{\pi}\int_{(M_A+M_B)^2}^{\infty} \frac{A_u(u',t)\,du'}{(u'-(M_A+M_B)^2+s_0-(M_A-M_B)^2-\eta+(t+\eta))(s_0-(M_A-M_B)^2-\eta-(t+\eta))}$$

we see that we will be able to use our previous arguments, once we have shown that the kernels have the property (P) for $s', u' > (M_A+M_B)^2+\varepsilon$.

This is true if

$$g_1(t) = \left[s_0-(M_A-M_B)^2-\eta-(t+\eta)\right]^{-1},$$

$$g_2(t) = g_1(t)\left[u'-(M_A+M_B)^2+s_0-(M_A-M_B)^2-\eta+(t+\eta)\right]^{-1}$$

have the property (P). Now since $s_0-(M_A-M_B)^2-\eta>0$ the Taylor series for $g_1(t)$ has the form

$$g_1(t) = \sum c_n^1 (t+\eta)^n \quad ; \quad -\eta \le t \le 0$$

with $c_n^1 \geq 0$ which proves (P) for $g_1(t)$.

Because $u' - (M_A + M_B)^2 > \varepsilon$ it is easy to see that $g_2(t)$ is of the form $[\alpha - \beta(t+\eta) - \gamma(t+\eta)^2]^{-1}$ with $\alpha, \beta, \gamma > 0$. The Taylor series for $g_2(t)$ then also has the form

$$g_2(t) = \sum c_n^2 (t+\eta)^n \quad ; \; -\eta \leq t \leq 0$$

with $c_n^2 \geq 0$ proving (P) for $g_2(t)$.

In V (20) we may therefore differentiate at $t = 0$ under the integral. This gives $\left(\frac{d}{dt}\right)^n \Phi(s_0, 0)$ as a sum of positive terms. In particular we obtain the inequalities

$$\frac{1}{s_0 - (M_A - M_B)^2 - 2\eta} \; \frac{1}{\pi} \int_{(M_A+M_B)^2}^{\infty} \frac{\left(\frac{d}{dt}\right)^n A_s(s',0)}{(s' - s_0)} ds' \leq \left(\frac{d}{dt}\right)^n \Phi(s_0, 0),$$

$$\frac{1}{s_0 - (M_A - M_B)^2 - 2\eta} \; \frac{1}{\pi} \int_{(M_A+M_B)^2}^{\infty} \frac{\left(\frac{d}{dt}\right)^n A_u(u',0)}{(u' - u_0)} du' \leq \left(\frac{d}{dt}\right)^n \Phi(s_0, 0)$$

$$u_0 = 2M_A^2 + 2M_B^2 - s_0 .$$

Therefore

$$F_R(s,t) \underset{def}{=} \frac{1}{\pi} \sum_{n=0}^{\infty} \frac{t^n}{n!} \int_{(M_A+M_B)^2}^{\infty} \frac{\left(\frac{d}{dt}\right)^n A_s(s',0)}{(s' - s)} ds',$$

V (21)

$$F_L(s,t) \underset{def}{=} \frac{1}{\pi} \sum_{n=0}^{\infty} \frac{t^n}{n!} \int_{(M_A+M_B)^2}^{\infty} \frac{\left(\frac{d}{dt}\right)^n A_u(u',0)\, du'}{(u' - u)},$$

$$u = 2M_A^2 + 2M_B^2 - s - t$$

satisfy the estimates

V (22)

$$|F_R(s,t)| \leq \mu(s_0, s)(s_0 - (M_A - M_B)^2 - 2\eta) \sum_{n=0}^{\infty} \frac{|t|^n}{n!} \left(\frac{d}{dt}\right)^n \Phi(s_0, 0)$$

$$\leq \mu(s_0, s)(s_0 - (M_A - M_B)^2 - 2\eta)\, \Phi(s_0, |t|),$$

$$|\mathcal{F}_L(s,t)| \le \mu(u_0,u)(s_0-(M_A-M_B)^2-2\eta)\,\Phi(s_0,|t|).$$

If we define $R'(s_0) = \text{Min}\,(R(s_0),\,s_0-(M_A-M_B)^2-2\eta)$ then $\Phi(s_0,t)$ is analytic in t for $|t| < R'(s_0)$.

But then also V (22) implies that

$$\mathcal{F}_R(s,t) + \mathcal{F}_L(s,t)$$

is analytic in

$$\tilde{\mathcal{C}}\,(R'(s_0)) = \left\{(s,t)\,\Big|\,|t| < R'(s_0),\,s \in \text{cut plane with cuts}\,(M_A+M_B)^2 \le (s,u) < \infty\right\}.$$

We want to show that this function is an analytic extension of $\mathcal{F}(s,t)$. Comparing the Taylor coefficients in the power series in t for both functions at s_0 we see that

$$\mathcal{F}(s_0,t) = \mathcal{F}_R(s_0,t) + \mathcal{F}_L(s_0,t)$$

But since we also may vary s_0 in a small interval, we indeed have

$$\mathcal{F}(s,t) = \mathcal{F}_R(s,t) + \mathcal{F}_L(s,t)$$

with the estimate

$$|\mathcal{F}(s,t)| \le \big(\mu(s_0,s) + \mu(u_0,u)\big)(s_0-(M_A-M_B)^2-2\eta)\,\Phi(s_0,|t|)$$

Repeating the arguments used above we may again extend the dispersion relation V (19) to all $(s,t) \in \tilde{\mathcal{C}}(R'(s_0))$ with $A_s(s',t)$ and $A_u(u',t)$ defined as in V (14). In particular, we do not need a subtraction.

Finally we are left with the case of a subtracted dispersion relation for $t_0 \le t \le 0$:

$$\mathcal{F}(s_0,t) - \sum_{n=0}^{N-1} \frac{(s_0-s_1)^n}{n!}\left(\frac{d}{ds}\right)^n \mathcal{F}(s_1,t) \quad =$$

$$= \frac{(s_0 - s_1)^N}{\pi} \int\limits_{(M_A+M_B)^2}^{\infty} \frac{A_s(s',t)}{(s'-s_1)^N (s'-s_0)} ds'$$

$$+ \frac{(u_0 - u_1)^N}{\pi} \int\limits_{(M_A+M_B)^2}^{\infty} \frac{A_u(u',t)}{(u'-u_1)^N (u'-u_0)} du',$$

$$u_0 + s_0 = u_1 + s_1 = 2M_A^2 + 2M_B^2 - t \, ; \; s_1 < s_0 \, ; \; t_0 \leq t \leq 0 .$$

Here we meet some trouble: If N is odd, we will not be able to apply the positivity property, since $u_0 - u_1 = s_1 - s_0 < 0$. If N is even, and we may always arrive at this situation, then we may essentially repeat the above arguments after multiplication by

$$(s_0 - (M_A - M_B)^2 - 2\eta - t)(s_1 - (M_A - M_B)^2 - 2\eta - t)^{-N}.$$

To conclude, we have the following situation concerning the question of subtractions: If the number of subtractions is even for $t_0 \leq t \leq 0$, it remains the same for $|t| < R'(s_0)$. If it is odd, we are forced to make an additional subtraction in order to apply the positivity properties and the number of subtractions may change by one. This fact was known a long time ago before this analyticity domain was discovered [J 2].

In the "Regge Pole" language this may be seen as follows: The Pomeranchuk trajectory has an even signature. So when $\alpha(t)$ goes through $2n$, this corresponds to a particle pole. From the fact that we have analyticity in $|t| < R'(s_0)$ we deduce that $\alpha(t)$ does not cross any even value for $0 < t < R'(s_0)$.
So we have the following implications:

$$\left.\begin{array}{lll} \alpha(0) < 0 & \Longrightarrow & \alpha(t) < 0 \\ \alpha(0) < 1 & \Longrightarrow & \alpha(t) < 2 \\ \alpha(0) < 2 & \Longrightarrow & \alpha(t) < 2 \end{array}\right\} \text{ for } 0 < t < R'(s_0).$$

Thus the case may appear that $\alpha(0) < 1$ and $\alpha(t) > 1$ for $0 < t < R'(s_0)$, i.e. starting with one subtraction, we may arrive at two subtractions for some positive $t < R'(s_0)$.

We mentioned at the beginning that the existence of a fixed t dispersion relation was an unessential assumption. To prove this let us start with the analyticity domain in s for $t \leq 0$ obtained by Bros, Epstein and Glaser [B 11].

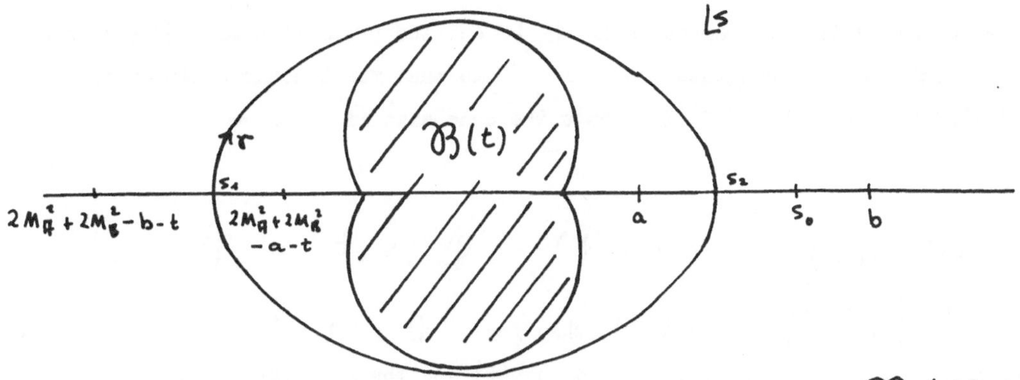

We have analyticity in s except for the real axis and the region $\mathcal{B}(t)$. Let t vary in an interval $t_0' \leq t \leq 0$, where t_0' will be fixed later.

Choose a,b sufficiently large such that

$$(M_A + M_B)^2 < a < b < \infty$$

$$s = a \notin \bigcup_{t_0' \leq t' \leq 0} \mathcal{B}(t') \; ; \; s = 2M_A^2 + 2M_B^2 - a - t \notin \bigcup_{t_0' \leq t' \leq 0} \mathcal{B}(t').$$

Define

$$\Phi_1(s,t) = \frac{1}{\pi} \int_a^b \frac{A_s(s',t)}{s'-s}\, ds',$$

$$\Phi_2(u,t) = \frac{1}{\pi} \int_a^b \frac{A_u(u',t)}{u'-u}\, du',$$

$$u = 2M_A^2 + 2M_B^2 - s - t.$$

It is clear that $\Phi_1(s,t)$ is analytic in (s,t) in the s-plane cut along $\overline{C}_{a,b}$,

$$C_{a,b} = \{ s \mid a < s < b \},$$

and the intersection for $s' \in \overline{C}_{a,b}$ of all fixed s' analyticity domains of $A_s(s',t)$ in t, i.e. in the intersection of the large Lehmann ellipses $E_L(s')$. That is, $\Phi_1(s,t)$ is analytic in

$$\{ (s,t) \mid s \in \overline{C}_{a,b} \, , \, t \in \bigcap_{a \leq s' \leq b} E_L(s') \}.$$

It may also be proved that the boundary values $\lim_{\varepsilon \to 0^+} \Phi_1(s \pm i\varepsilon, t)$ for $s \in C_{a,b}$

have the same analyticity properties in t [S 4]. The properties of $\Phi_2(u,t)$ are analogous with u substituted for s (Note that the Lehmann ellipses for $A_u(u',t)$ and $A_s(s',t)$ have the same size if $u' = s'$).

Consider

$$\Phi(s,t) = F(s,t) - \Phi_1(s,t) - \Phi_2(u,t)$$

If $[t_0', 0] \subset \bigcap_{a \le s' \le b} E_L(s')$ then for $t \in [t_0', 0]$, $\Phi(s,t)$ is analytic in s except for $s \in \mathcal{B}(t)$ and the real intervals

$(-\infty, 2M_A^2 + 2M_B^2 - b - t]$,

$[2M_A^2 + 2M_B^2 - a - t, a]$ and $[b, \infty)$.

Choose a loop γ enclosing $\bigcup_{t_0' \le t \le 0} \mathcal{B}(t)$ in the clockwise sense such that it intersects the real axis once on the intervals $[a, b]$ and $[2M_A^2 + 2M_B^2 - b - t, 2M_A^2 + 2M_B^2 - a - t]$ for all $t \in [t_0', 0]$. It is not hard to prove the existence of $t_0' < 0$, a, b and γ satisfying all the conditions above.

But then, for $t \in [t_0', 0]$

$$\Psi(s,t) = \Phi(s,t) - \frac{1}{2\pi i} \int_\gamma \frac{\Phi(s',t)}{s'-s}\, ds'$$

is an analytic function in the cut s -plane with cuts

$$-\infty < s \le 2M_A^2 + 2M_B^2 - b - t \ ; \quad b \le s < \infty .$$

This follows from Laurent's theorem.

With $\Psi(s,t)$ we may now repeat all the preceeding arguments. All we have to do is to show the existence of a real point s_0 between the cuts such that $\Psi(s,t)$ is analytic in a neighborhood of $s = s_0$, $t = 0$.

Choose $s_0 \in [a, b]$ such that s_0 lies outside the region enclosed by γ. Then there exists, according to Bros, Epstein and Glaser [B 10], a neighborhood of s_0 and $t = 0$, such that $F(s,t)$ is analytic in s and t except for the physical cut in s. If we subtract $\Phi_1(s,t)$, then the cut disappears. This is so since $F(\bar{s}_0 + i0, t)$ and $F(\bar{s}_0 - i0, t)$ $(-\varepsilon + s_0 \le \bar{s}_0 \le s_0 + \varepsilon)$ are analytic in t because of the existence of a small Lehmann ellipse. But the discontinuity of $F(s,t) - \Phi_1(s,t)$ across the cut at $\bar{s}_0$ is just $F(\bar{s}_0 + i0, t) - F(\bar{s}_0 - i0, t) - 2i A_s(\bar{s}_0, t)$ which vanishes for $t \le 0$, $|t|$

sufficiently small by definition of $A_s(\bar{s}_0, t)$. Therefore it vanishes for all t in a neighborhood of $t = 0$.
Since $\Phi_2(u,t)$ is analytic in a neighborhood of $s = s_0, t = 0$ the same holds for $\Phi(s,t)$.

Furthermore the results of Bros, Epstein and Glaser may be extended in such a way that to each point $\tilde{s}$ in the cut plane (with cuts $(M_A + M_B)^2 \leq s < \infty$, $-\infty < s \leq (M_A - M_B)^2$), there exists a neighborhood of $s = \tilde{s}, t = 0$ where $F'(s,t)$ is analytic [R 1] .

This property of $F(s,t)$ holds in particular for all non real points $s \in \gamma$ and therefore also for $\Phi(s,t)$. But for the two real points $s_1, s_2 \in \gamma$ $\Phi(s,t)$ is still analytic in a neighborhood of $s = s_1$ (resp. $s = s_2$), $t = 0$ if we repeat the argument given above for $s = s_0$. Using the compactness of γ we have thus shown the existence of a neighborhood of $t = 0$, such that $\Phi(s',t)$ is analytic in t there for all $s' \in \gamma$.

Since $s' - s$ never vanishes if $s' \in \gamma$ and if s varies in a neighborhood of s_0 , this proves the analyticity of $\psi(s,t)$ in a neighborhood of $s = s_0, t = 0$.
Finally we may remark that the last discussion also covers the case where unphysical thresholds appear in the dispersion relations and therefore where on a part of the s- and/or u-cut the absorptive part no longer has the positivity properties.

VI. Quantitative estimate of the domain of validity of dispersion relations

We return to the case where we have dispersion relations with normal thresholds. Having obtained analyticity for $F'(s,t)$ in $\mathcal{E}(R'(s_0))$ it is a natural question to ask how large $R'(s_0)$ in fact may become. Let us see what we may expect: If we look at the non-relativistic potential scattering given by a superposition of Yukawa potentials:

$$r\,V(r) = \int e^{-mr}\, d\sigma(m),$$

$$\mathrm{supp}\ \sigma \subset [m_0, \infty)\ ;\ m_0 > 0,$$

then from the Mandelstam representation, which is valid in that case, we obtain the condition $|t| < 4m_0^2$ [B 4] . Thus $R'(s_0)$ should essentially be determined by the mass of the lightest particle entering the theory.

Now unfortunately the work of Bros et al. does not provide any numerical information for $R'(s_0)$. For physical s , however, we have analyticity of $F'(s,t)$ in t in the small Lehmann ellipse $E_{sL}(s)$, while the absorptive part is analytic in the large Lehmann ellipse $E_L(s)$.
In order to take advantage of this information, the following trick was used by Sommer [S 3]

Cosider

$$\text{VI (1)} \qquad F^{s_1}(s,t) = F(s,t) - \frac{1}{\pi} \int_{(M_A+M_B)^2}^{s_1} \frac{A_s(s',t)}{(s'-s)} ds' .$$

For $t_0 \leq t \leq 0$, $F^{s_1}(s,t)$ has a right hand cut starting at s_1 instead of $(M_A + M_B)^2$ and has the same positivity properties as $F(s,t)$. Consider $s = s_0$ in the interval $((M_A + M_B)^2, s_1)$. Then the right hand side of VI (1) is analytic in t in

$$E_{sL}(s_0) \cap \left(\bigcap_{(M_A+M_B)^2 \leq s' \leq s_0} E_L(s') \right)$$ (see page 31).

If we define $r_L(s)$ (resp $r_{sL}(s)$) to be the right extremity of $E_L(s)$ (resp. $E_{sL}(s)$), then in particular $F^{s_1}(s_0,t)$ is analytic in

$$|t| < \operatorname{Min}\left(r_{sL}(s_0), \operatorname{Min}_{(M_A+M_B)^2 \leq s' \leq s_1} r_L(s')\right) \underset{def.}{=} R(s_0, s_1).$$

The function $r_L(s)$ is given by

$$r_L(s) = 4 \frac{(M_A'^2 - M_A^2)(M_B'^2 - M_B^2)}{s - (M_A' - M_B')^2}$$

(c.f. IV (2)),

and is obviously monotonically decreasing in s. Therefore

$$R(s_0, s_1) = \operatorname{Min}\left(r_{sL}(s_0), r_L(s_1)\right).$$

Since the large Lehmann ellipse is always strictly larger than the small one at the same energy, we can choose s_0 and s_1 sufficiently close to each other so that

$$R(s_0, s_1) = r_{sL}(s_0).$$

Let s_{max} be the point where $r_{sL}(s)$ attains its maximum:

$$r_{sL}(s_{max}) = \operatorname{Max}_{(M_A+M_B)^2 < s < \infty} r_{sL}(s).$$

Then starting from $F^{s_1}(s_{max}, t)$ (s_1 sufficiently close to s_{max}, $s_1 > s_{max}$) we obtain, repeating the by now well known construction, a function analytic in

$$\{(s,t) \mid |t| < r_{sL}(s_{max}),\ s \notin [s_1, \infty),\ u \notin [(M_A + M_B)^2, \infty)\}$$

That this function is an analytic continuation of $F^{s_1}(s,t)$ is easily seen taking into account the fact that $F^{s_1}(s,t)$ is analytic in (s,t) in a neighborhood of $s = s_0$ and $t = 0$ (see page 17). But then $F(s,t)$ is analytic in $\mathcal{E}(r_{sL}(s_{max}))$ because of the analyticity of

$$\frac{1}{\pi} \int_{(M_A+M_B)^2}^{s_1} \frac{A(s',t)}{(s'-s)} ds'$$

(see page 31).

Let us now inspect

$$R = r_{sL}(s_{max}) = \underset{(M_A+M_B)^2 \leq s < \infty}{Max} 2k^2 \left\{ \left[1 + \frac{(M_A'^2 - M_A^2)(M_B'^2 - M_B^2)}{k^2(s - (M_A' - M_B')^2)} \right]^{\frac{1}{2}} - 1 \right\},$$

$$k^2 = \frac{(s - (M_A - M_B)^2)(s - (M_A + M_B)^2)}{4s}.$$

If

$$\Delta = M_A' - M_A = M_B' - M_B,$$

then

$$s_{max} = (M_A + M_B)(M_A + M_B + \Delta),$$

$$R = (M_A' - M_A)^2.$$

In many cases $\Delta = 2\mu$ (e.g. $\langle 0 | j_\pi(0) | 3\pi \rangle \neq 0$, $\langle 0 | j_K(0) | K 2\pi \rangle \neq 0$) and we have $R = 4\mu^2$. This is the best possible result since $t = 4\mu^2$ is a singular point.

Note also that $s_{max} = 8\mu^2$ for pion-pion scattering is exactly the point where the large Lehmann ellipse with right extremity $r_L(s) = 256\mu^4 \cdot s^{-1}$ touches the border line $t = 16\mu^2 + 64\mu^4(s - 4\mu^2)^{-1}$ of the support of the Mandelstam spectral function ρ_{st}.

For elastic KK - and $K\bar{K}$ - scattering, it is also possible to show that $R = 4\mu^2$.

The difficulty that arises in the $K\,K$ scattering for example is that the left hand cut starts at $u_{thr} = 4\mu^2$ instead of $u_{thr} = 4M_K^2$ because of the two pion threshold in the $K\bar{K}$ channel. Since the absorptive part is not positive there, the contribution to the amplitude coming from this part has to be subtracted in order to apply the above arguments (see page 33). For pion-nucleon scattering $\langle 0 | j_N(0) | N\pi \rangle \neq 0$ and therefore $\mu^2 < R < 4\mu^2$. A similar situation appears for the pion-lambda scattering. To discuss these cases Bessis and Glaser [B 3] returned to the initial method and computed $R'(s_0)$ for $s_0 \in [(M-\mu)^2, (M+\mu)^2]$ where M is the mass of the heavy particles, and found $R'(s_0)$ to be $4\mu^2$ in agreement with the expectations.

VII Bounds on the amplitudes and cross sections

As an application of these results let us prove the Froissart bound for the cross-section. First we note that due to V (14) $A_s(s,t)$ is holomorphic in $|t| < R$ for all s. The analyticity in $E_L(s)$ implies the convergence of the Legendre series

$$A_s(s,t) = \frac{s^{\frac{1}{2}}}{2k} \sum_{\ell=0}^{\infty} (2\ell+1)\, \operatorname{Im} f_\ell(s)\, P_\ell\left(1 + \frac{t}{2k^2}\right)$$

in $E_L(s)$. From III (3') we also have $\operatorname{Im} f_\ell(s) \geq 0$. But then $A_s(s,t)$ is analytic in an ellipse with foci $0, -4k^2$ and right extremity R (see Appendix). This ellipse is called the unitarity ellipse $E_U(R,s)$. Now we look for the largest value of $A_s(s,0)$ which is compatible with unitarity (III(3')) and a given value of $A_s(s, t=R)$. Since $P_\ell(x)$ is increasing in ℓ for $x>1$ is is clear that the distribution of the $\operatorname{Im} f_\ell(s)$ maximizing $A_s(s,t=0)$ is given by

$$\begin{aligned} \operatorname{Im} f_\ell(s) &= 1 & & 0 \leq \ell \leq L(s)-1 \\ \operatorname{Im} f_\ell(s) &= \varrho(s) & & \ell = L(s)\,;\; 0 \leq \varrho(s) < 1 \\ \operatorname{Im} f_\ell(s) &= 0 & & \text{other wise} \end{aligned}$$

such that

$$A_s(s, t=R) = \frac{s^{\frac{1}{2}}}{2k} \left\{ \sum_{\ell=0}^{L(s)-1} (2\ell+1)\, P_\ell\left(1+\frac{R}{2k^2}\right) + (2L(s)+1)\varrho(s) \cdot P_{L(s)}\left(1+\frac{R}{2k^2}\right) \right\}$$

We then have

$$\frac{2k}{s^{\frac{1}{2}}} A_s(s,0) = \sum_{\ell=0}^{L(s)-1} (2\ell+1) + \varrho(s)\,(2L(s)+1) < (L(s)+1)^2,$$

which gives

VII(2)
$$\sigma_{tot}(s) < \frac{4\pi}{k^2}\,(L(s)+1)^2,$$
$$\frac{L(s)}{k} \geqslant \sqrt{\frac{\sigma_{tot}(s)}{4\pi}} - \frac{1}{k}.$$

Now we want to estimate $L(s)$ as a function of $A_s(s, t = R)$.

Laplace's representation of the Legendre polynomials gives

$$P_\ell(x) = \frac{1}{\pi}\int_0^{\pi} \left(x + \cos\phi\sqrt{x^2-1}\right)^\ell d\phi \geq \frac{1}{\pi}\int_0^{\frac{\pi}{2}} \left(x + \cos\phi\sqrt{x^2-1}\right)^\ell d\phi$$

$$\geqslant \frac{\phi_0}{\pi}\left(x + \cos\phi_0\sqrt{x^2-1}\right)^\ell \;;\quad 0 \leqslant \phi_0 \leqslant \frac{\pi}{2} \;;\quad x > 1.$$

Setting $y = \left(x + \cos\phi\sqrt{x^2-1}\right)$, we obtain

VII(3)
$$\sum_{\ell=0}^{L} (2\ell+1) P_\ell(x) \geqslant 2\sum_{\ell=0}^{L} \ell P_\ell(x) \geqslant \frac{2\phi_0}{\pi}\sum_{\ell=0}^{L} \ell y^\ell$$

$$= \frac{2\phi_0}{\pi}\, y\,\frac{d}{dy}\sum_{\ell=0}^{L} y^\ell = \frac{2\phi_0}{\pi}\, y\left\{\frac{(L+1)y}{(y-1)} - \frac{y^{L+1}-1}{(y-1)^2}\right\}.$$

Choosing in particular

$$x = 1 + \frac{R}{2k^2} \;;\quad L = L(s) - 1$$

we may distinguish two cases

(a) $\frac{L(s)}{k}$ is bounded for $s \to \infty$.

But then $\sigma_{tot}(s)$ also remains bounded.

(b) $\frac{L(s)}{k} \to \infty$ for $s \to \infty$.

Then we may neglect the second term in VII(3) and obtain, using $y - 1 \approx \cos\phi_0 \frac{R^{\frac{1}{2}}}{k}$ for $s \to \infty$:

$$\text{VII(4)} \quad \frac{A_s(s, t=R)}{s} > C(\phi_0)\, \frac{L(s)}{k}\, R^{\frac{1}{2}} \left(1 + \cos\phi_0 \frac{R^{\frac{1}{2}}}{k}\right)^{L(s)}$$

for sufficiently large s. Note that ϕ_0 may be chosen arbitrary in $[0, \frac{\pi}{2}]$ and that $C(\phi_0) \to 0$ for $\phi_0 \to 0$. We may therefore write VII(4) as

$$\text{VII(5)} \quad \frac{A_s(s, t=R)}{s} > C'(\phi_0)\, \frac{L(s)}{k}\, R^{\frac{1}{2}}\, \exp\left\{\cos\phi_0 \cdot \frac{R^{\frac{1}{2}}}{k} \cdot L(s)\right\}$$

with $C'(\phi_0) \to 0$ for $\phi_0 \to 0$ and $s > s(\phi_0)$.

We assumed $F(s, t)$ to satisfy a subtracted dispersion relation. But then

$$\int_{s_{thr}}^{\infty} \frac{A_s(s, t=R)}{s^{n+1}}\, ds$$

should converge for suitable n, i.e. $A_s(s, t=R)$ should be "polynomially bounded in measure". Combined with VII(2) and VII(5) this gives the condition

$$\text{VII(6)} \quad C'(\phi_0) \int_{s(\phi_0)}^{\infty} s^{-n-1} \left(\sqrt{\frac{\sigma_{tot}(s)}{4\pi}} - \frac{1}{k}\right) \exp\left\{\cos\phi_0\, R^{\frac{1}{2}} \left(\sqrt{\frac{\sigma_{tot}(s)}{4\pi}} - \frac{1}{k}\right)\right\} ds < \infty$$

This is an exact but not very handy condition. Note that this condition is satisfied if we have case (a). However, if we may write $A_s(s, t=R) < s^N$, then

$$\text{VII(7)} \quad s^{N-1} > C'(\phi_0)\, \frac{L(s)}{k}\, R^{\frac{1}{2}}\, \exp\left\{\cos\phi_0\, R^{\frac{1}{2}}\, \frac{L(s)}{k}\right\}$$

and we have

$$\cos\phi_0\, \frac{L(s)}{k}\, R^{\frac{1}{2}} \leq (N-1) \log s$$

for $\phi_0 > 0$, which gives

$$\sigma_{tot}(s) < \frac{4\pi}{R\cos^2\phi_0} (N-1)^2 (\log s)^2$$

which is Froissart's bound [F 1] .

Next we want to obtain a bound for the scattering amplitude. We have

$$\text{VII(8)} \qquad \frac{2k}{s^{\frac{1}{2}}} F(s,0) = \sum_{\ell=0}^{\infty} (2\ell+1) f_\ell(s)$$

and we look for a maximalization for given $A_s(s, t = R)$.

We assume $s^N > A_s(s, t = R)$.

Since

$$\frac{2k}{s^{\frac{1}{2}}} A_s(s, t = R) = \sum_{\ell=0}^{\infty} (2\ell+1) \, \mathrm{Im}\, f_\ell(s) P_\ell\left(1 + \frac{R}{2k^2}\right)$$

is a sum of positive terms, each term must be less than the sum. The inequality $|f_\ell(s)|^2 < \mathrm{Im}\, f_\ell(s)$ then gives

$$\text{VII(9)} \qquad |f_\ell(s)| \leq \frac{s^{\frac{N}{2}}}{\left\{(2\ell+1) P_\ell\left(1 + \frac{R}{2k^2}\right)\right\}^{\frac{1}{2}}} \leq \frac{s^{\frac{N}{2}}}{\left\{(2\ell+1) \frac{\phi_0}{\pi} \left(1 + \cos\phi_0 \frac{R^{\frac{1}{2}}}{k}\right)^{\ell}\right\}^{\frac{1}{2}}}$$

Eq. VII(9) is an improvement of the unitarity condition $|f_\ell(s)|^2 \leq 1$ if $\ell \approx k \log s$. The contributions to VII(8) for $\ell \gg k \log s$ then become negligible and we have

$$|F(s,0)| < \text{const } s(\log s)^2$$

A more careful discussion gives the optimum result for the partial wave distribution [L 5]

$$f_\ell(s) = 1 \quad ; \quad \ell = 0, 1, \dots \hat{L}(s) ;$$

$$f_\ell(s) = \frac{c(s)}{P_\ell\left(1 + \frac{R}{2k^2}\right)} ; \quad \ell > \hat{L}(s) ;$$

where

$$1 < P_{\hat{L}(s)}\left(1 + \frac{R}{2k^2}\right) < C(s) \leq P_{\hat{L}(s)+1}\left(1 + \frac{R}{2k^2}\right),$$

such that

$$|F(s,0)| < \frac{s\, N^2 (\log s)^2}{4 R \cos^2\phi_0}$$

By crossing symmetry, $|F(s,0)|$ must be bounded by const $\cdot |s| (\log |s|)^2$ on both the right and left hand cut. Since $F(s,0)$ is polynomially bounded in the cut plane, the Phragmen-Lindelöf theorem then implies that

$$|F(s,0)| < \text{const.}\, |s| \, (\log |s|)^2$$

in the cut plane.

Therefore we need <u>at most two subtractions </u>for $t = 0$.

If we use the positivity property $|A_s(s,t)|_{-4k^2 \leq t \leq 0} < A_s(s,0)$ it is easy to prove that this also must hold for $t_0 \leq t \leq 0$. Since for $|t| < R$ the (even) number of subtractions is conserved, we have proved the validity of two subtractions in $\tilde{\mathcal{E}}(R)$. Thus we may choose $N = 2$ in the above considerations.

For $R = 4\mu^2$ we get

$$\sigma_{tot}(s) \leq \frac{\pi}{\mu^2 \cos^2 \phi_0} (\log s)^2$$

which should hold in the mean asymptotically for s . Note that $\pi \mu^{-2} = 60$ millibarn, which is very reasonable.

The interesting information, however, lies in VII(6) if we put $n = 2$.

For π-π- scattering it is possible to obtain a numerical bound for the integral VII(6) , as will be shown in section XIII.

In order to estimate away from the forward direction, we use

$$|P_\ell(\cos\theta)| < 2\sqrt{\frac{1}{\pi(2\ell+1)\sin\theta}} \quad ; \quad 0 < \theta < \pi .$$

A repetition of the above discussion then gives

$$\text{VII(10)} \qquad |A_s(s,\cos\theta)| \leq |F(s,\cos\theta)| \leq C \, \frac{s^{\frac{3}{4}} (\log s)^{\frac{3}{2}}}{(\sin\theta)^{\frac{1}{2}}} .$$

This estimate may be improved to [K 3] :

$$|A_s(s,\cos\theta)| \leq |F(s,\cos\theta)| \leq C \, \frac{(\log s)^{\frac{3}{2}}}{\sin^2\theta}$$

such that the fixed angle elastic differential cross-section goes to zero at high energies

$$\text{VII(11)} \qquad \frac{d\sigma_{el}}{d\Omega} \leq C \, \frac{C\,(\log s)^3}{s \cdot \sin^4\theta} \quad ; \quad 0 < \theta < \pi .$$

VIII Inelastic processes

Up till now we have only considered elastic two-body amplitudes. The principal ingredient for finding the above results was the positivity property of the absorptive part. For inelastic two-body reactions, A + B $\longrightarrow$ C + D , a corresponding property does not hold. We may write the partial wave expansion for $F_{AB \to CD}(s,t)$ as follows [C 3] [S 7] :

$$F_{AB \to CD}(s,t) = a(s) \sum_{\ell=0}^{\infty} (2\ell+1)\,(CD|T|AB)_\ell(s)\, P_\ell\left(\frac{k_{AB}^2(s) + k_{CD}^2(s) + \frac{\delta}{s} + t}{2\,k_{AB}(s)\,k_{CD}(s)}\right)$$

where $a(s)$ is an unimportant kinematical factor and

$$\delta = -\frac{1}{4}\left(M_A^2 - M_B^2 - M_C^2 + M_D^2\right),$$

$$k_{AB}(s) = \frac{\left(s - (M_A - M_B)^2\right)^{\frac{1}{2}} \left(s - (M_A + M_B)^2\right)^{\frac{1}{2}}}{2\,s^{\frac{1}{2}}},$$

$$k_{CD}(s) = \frac{\left(s - (M_C - M_D)^2\right)^{\frac{1}{2}} \left(s - (M_C + M_D)^2\right)^{\frac{1}{2}}}{2\,s^{\frac{1}{2}}}.$$

Now as a consequence of unitarity, we have

$$\mathrm{Abs}\,(CD|T|AB)_\ell = \sum_n (CD|T^\dagger|n)_\ell\,(n|T|AB)_\ell,$$

where $\{n\}$ is a complete set of intermediate states.

This can be combined with Schwarz inequality:

$$\begin{aligned}
|\mathrm{Abs}\,(CD|T|AB)_\ell|^2 &= \Big|\sum_n (CD|T^\dagger|n)_\ell\,(n|T|AB)_\ell\Big|^2 \\
&\le \sum_n |(CD|T^\dagger|n)_\ell|^2 \sum_n |(n|T|AB)_\ell|^2 \\
&\le \mathrm{Abs}\,(CD|T|CD)_\ell \cdot \mathrm{Abs}\,(AB|T|AB)_\ell
\end{aligned}$$

where the last equality again follows from unitarity.

Thus

VIII (1) $\left| \mathrm{Abs}\,(CD|T|AB)_\ell(s) \right| \leq \left(\mathrm{Abs}\,(CD|T|CD)_\ell(s)\right)^{\frac{1}{2}} \left(\mathrm{Abs}\,(AB|T|AB)_\ell(s)\right)^{\frac{1}{2}}.$

Now we know that

$$A_s^{AB \to AB}(s,t) = \frac{s^{\frac{1}{2}}}{2k_{AB}} \sum_{\ell=0}^{\infty} (2\ell+1)\,\mathrm{Abs}\,(AB|T|AB)_\ell(s)\, P_\ell(\cos\theta_s)$$

[resp.

$$A_s^{CD \to CD}(s,t) = \frac{s^{\frac{1}{2}}}{2k_{CD}} \sum_{\ell=0}^{\infty} (2\ell+1)\,\mathrm{Abs}\,(CD|T|CD)_\ell(s)\, P_\ell(\cos\theta_s) \Big]$$

is analytic in an ellipse with foci ± 1 and semimajor axis

$$\left(1 + \frac{R_{AB}}{2k^2_{AB}}\right) \qquad \left[\text{resp. } \left(1 + \frac{R_{CD}}{2k^2_{CD}}\right) \right].$$

Therefore (VIII (1)) gives (see page 16):

$$\overline{\lim_\ell} \left| \mathrm{Abs}\,(CD|T|AB)_\ell(s) \right|^{\frac{1}{\ell}} \leq \frac{1}{\zeta(s)^{\frac{1}{2}}},$$

$$\zeta(s) = \left(1 + \frac{R_{AB}}{2k^2_{AB}} + \sqrt{\left(1+\frac{R_{AB}}{2k^2_{AB}}\right)^2 - 1}\right)\left(1 + \frac{R_{CD}}{2k^2_{CD}} + \sqrt{\left(1+\frac{R_{CD}}{2k^2_{CD}}\right)^2 - 1}\right).$$

Thus for fixed $s \geq s_{thr}$, $\mathrm{Abs}\, F_{AB \to CD}(s,t)$ becomes analytic in t in the ellipse

VIII (3) $\quad E'_v(R_A, R_B, s) =$

$$\left\{ t \,\middle|\, \left|t + (k_{AB} - k_{CD})^2 + \frac{\delta}{s}\right| + \left|t + (k_{AB} + k_{CD})^2 + \frac{\delta}{s}\right| < \frac{4k_{AB}\, k_{CD}\,(1+\zeta(s))}{2\,\zeta(s)^{\frac{1}{2}}} \right\}.$$

For sufficiently large s this ellipse takes the approximate form

$$\left\{ t \;\middle|\; |t| + |t + s - \sum M_i^2| < s - \sum M_i^2 + \frac{1}{2}\left(R_{AB}^{\frac{1}{2}} + R_{CD}^{\frac{1}{2}}\right)^2 \right\}$$

so that the right extremity tends to $\frac{1}{4}\left(R_{AB}^{\frac{1}{2}} + R_{CD}^{\frac{1}{2}}\right)^2$.

This has been applied to the $K\pi \to K\pi$ scattering [S 6] :

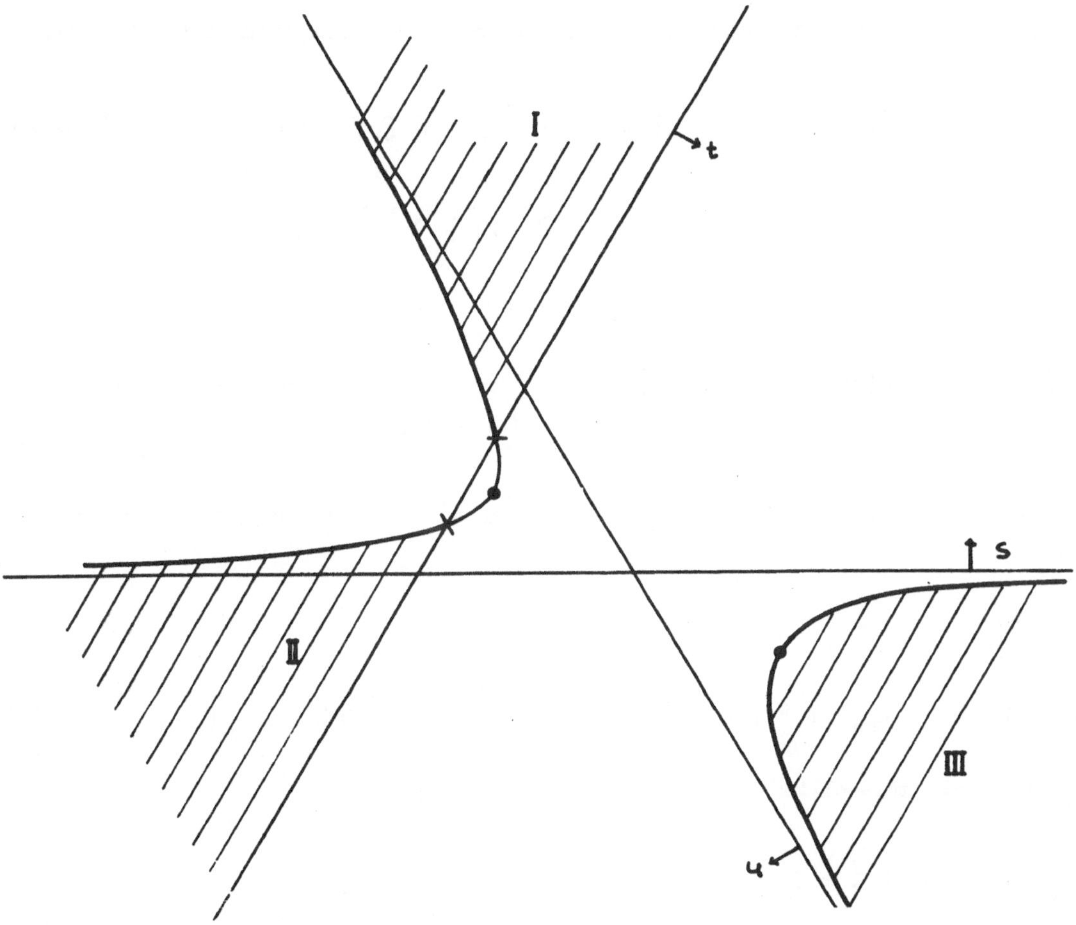

The physical regions are

I: $K + \pi \to K + \pi$

II: $K + \pi \to K + \pi$

III: $K + \bar{K} \to \pi + \pi$

Thus the positivity property may immediately be applied to prove fixed t dispersion relations in $|t| < 4\mu^2$. We may now ask whether fixed u dispersion relations can be proved. There the absorptive part of the $K + \bar{K} \to \pi + \pi$ scattering amplitude comes in. Now VIII (3) (with u substituted for t) gives us analytic properties in u, where $R_{\pi\pi} - R_{K\bar{K}} = 4\mu^2$. The analyticity properties of $A_s^{K+\pi \to K+\pi}(s,t)$ in u for fixed s are easily obtained from those in t: Note that the left focus of the unitarity ellipse with foci $-4k^2_{K\pi}$, 0 and right extremity $R_{K\pi} = 4\mu^2$ approaches the point $u = 0$ for $s \to \infty$.

Another application of the inequality (VIII (1)) has been made in the discussion of the reaction $\pi + \Lambda \to \pi + \Sigma$. For details we refer the reader to the original paper [S 7].

The second application of the unitarity condition for reaction amplitudes involves the inequality

$$\text{VIII (4)} \qquad \text{Abs}\,(AB|T|AB)_\ell = \sum_n (AB|T^+|n)_\ell (n|T|AB)_\ell \geq |(CD|T|AB)_\ell|^2 .$$

This implies

$$\overline{\lim_\ell}\, |(CD|T|AB)_\ell(s)|^{\frac{1}{\ell}} \leq \frac{1}{\left(1 + \frac{R_{AB}}{2k^2_{AB}} + \sqrt{\left(1 + \frac{R_{AB}}{2k^2_{AB}}\right)^2 - 1}\right)},$$

so that the scattering amplitude $T_{AB \to CD}(s,t)$ itself is analytic in t in the ellipse

$$\text{VIII (5)} \qquad E^2_U(R_{AB}, s) =$$

$$\left\{ t \,\middle|\, |t + (k_{AB} - k_{CD})^2 + \tfrac{\delta}{s}| + |t + (k_{AB} + k_{CD})^2 + \tfrac{\delta}{s}| < 4 k_{AB} k_{CD} \left(1 + \frac{R_{AB}}{4k^2_{AB}}\right)^{\frac{1}{2}} \right\}$$

For $s \to \infty$ this ellipse takes the approximate form

$$\left\{ t \,\middle|\, |t| + |t + s - \sum M_i^2| < s - \sum M_i^2 + \frac{R_{AB}}{2} \right\}$$

so that the right extremity tends to $\frac{R_{AB}}{4}$. These statements of course hold in the special case $A = C$, $B = D$.

As an example let us take the reaction $\pi\pi \to N\bar{N}$. We obtain analyticity in $|t| < \mu^2$ for sufficiently large s.

We finally turn to the discussion of multiple production amplitudes:

$$A + B \to C + n \qquad \text{(A,B,C scalar particles)}$$

where n denotes a certain system of particles.

Now such amplitudes depend oh a large set of variables. It is convenient to keep the energy s and the emission angle θ of the particle C as variables because in most cases summation is carried out over the other variables when differential cross sections are considered. In the case where n consists of two particles, it has been shown by Ascoli and Minguzzi[A2], using the Jost-Lehmann-Dyson representation, that $\frac{d}{d\Omega}\sigma(s,\cos\theta)$ is analytic in the small Lehmann ellipse corresponding to the process A + B $\rightarrow$ A + B. Note that we may write for real $\cos\theta$

$$|T(s,\cos\theta)|^2 = T^*(s,\cos\theta^*)\cdot T(s,\cos\theta)$$

and then we can continue the r.h.s. for comlex $\cos\theta$. This result has been extended by Logunov, Mestrivishvili and van Hieu[L3] and by Tiktopoulos and Treiman[T1] to the case where n is arbitrary and the Lehmann ellipse has been replaced by a larger one coming from positivity.

We start from unitarity

$$\mathrm{Abs}\,\langle AB|T|AB\rangle \geq \sum_n |\langle AB|T|C+n\rangle|^2$$

and write

$$T_{AB\to C+n}(s,\cos\theta,\alpha,\mu) = \sum_{\ell=0}^{\infty}(2\ell+1)\,a_\mu^\ell(s,\alpha)\,d_{\mu 0}^\ell(\theta).$$

Here

$$d_{\mu 0}^\ell = \langle \ell\mu|\exp -iJ_y\cdot\theta|\ell 0\rangle$$

are Legendre's associated functions of the first kind, μ describes the helicity of the compound n and α other variables necessary for the complete description of $T_{AB\to C+n}$. Then in our notation

$$\mathrm{Abs}\,(AB|T|AB)_\ell(s) \geq \oint_\alpha \sum_{\mu=-\ell}^{+\ell} |a_\mu^\ell(s,\alpha)|^2.$$

Let us look at

$$\oint_\alpha \sum_\mu |T_{AB\to C+n}(s,\cos\theta,\alpha,\mu)|^2 = \sum_\ell\sum_{\ell'}\oint_\alpha\sum_\mu (2\ell+1)(2\ell'+1)\,a_\mu^{\ell *}(s,\alpha)\,a_\mu^{\ell'}(s,\alpha)\cdot$$
$$\cdot\, d_{\mu 0}^\ell(\theta)\,d_\mu^{\ell'}(\theta).$$

Now we may use the Schwarz inequality and unitarity again

$$\left| \oint_\alpha \sum_\mu a_\mu^{\ell *}(s,\alpha) a_\mu^{\ell'}(s,\alpha) \right| \leq \left(\oint_\alpha \sum_\mu |a_\mu^{\ell}(s,\alpha)|^2 \right)^{\frac{1}{2}} \left(\oint_\alpha \sum_\mu |a_\mu^{\ell'}(s,\alpha)|^2 \right)^{\frac{1}{2}}$$

$$\leq \left(\mathrm{Abs}(AB|T|AB)_\ell(s) \right)^{\frac{1}{2}} \left(\mathrm{Abs}(AB|T|AB)_{\ell'}(s) \right)^{\frac{1}{2}}$$

to obtain

$$\left| \oint_\alpha \sum_\mu a_\mu^{\ell *}(s,\alpha)\, a_\mu^{\ell'}(s,\alpha) \right| \leq \frac{C(s)}{\left(\cos\Theta(s) + (\cos^2\Theta(s) - 1)^{\frac{1}{2}}\right)^{\frac{\ell+\ell'}{2}}}$$

where $\cos\Theta(s) = 1 + \frac{R_{AB}}{2k^2_{AB}}$ is given by the analyticity of $A_s^{AB \to AB}(s,t)$ in the unitarity ellipse. Now it is shown in the Appendix that

VIII (6) $$|d^{\ell}_{\mu 0}(\Theta)| < (x + \sqrt{x^2 - 1})^{\ell}$$

if $\cos\Theta$ varies in an ellipse with foci ± 1 and semimajor axis $x > 1$.
We have therefore:

$$\oint_\alpha \sum_\mu |\Psi_{AB \to C+u}(s, \cos\Theta, \alpha, \mu)|^2$$

is analytic in an ellipse with foci ± 1 and semimajor axis

$\cos\bar{\Theta}(s) = \left(1 + \frac{R_{AB}}{4k^2_{AB}}\right)^{\frac{1}{2}}$. Expressed in the variable $t = (q_A - q_C)^2$ we obtain an ellipse, whose right extremity tends to $\frac{R_{AB}}{4}$. In particular, it is possible to repeat the arguments leading to the Froissart bound. In this argument it is not necessary to integrate over all the allowed intervals in the variables associated with the particles.

As an interesting example let us consider the "single pion exchange model" for the scattering $N + \pi \to N + \pi + \pi$, taking N to be the particle C in the discussion above [for a discussion of this model see [K 1] and the literature quoted there] :

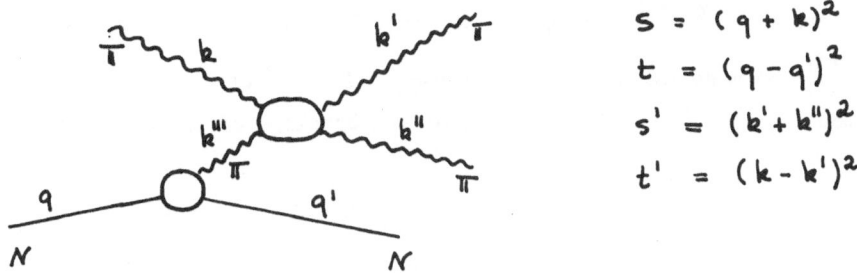

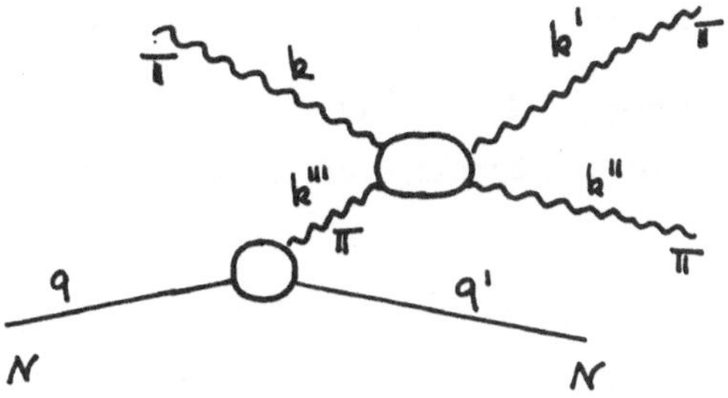

$$s = (q+k)^2$$
$$t = (q-q')^2$$
$$s' = (k'+k'')^2$$
$$t' = (k-k')^2$$

$\Psi(s',t';t)$ is an off shell scattering amplitude due to the fact that $t = k'''^2 \neq \mu^2$. $\Gamma(t)$ is the pion-nucleon vertex. In order to determine the contribution of the above graph to the full amplitude and to get information on the elastic $\pi-\pi$ scattering amplitude, one would like to extend $\Psi_{N\pi \to N\pi\pi}(s,t)$ to $t = \mu^2$. This is possible in the limit of very high energies due to the above discussion since $R_{N\pi} = 4\mu^2$. Also $\Gamma(\mu^2)$ is determined by pion-nucleon scattering experiments. Note that the interesting interval for s' is $750 \leq s' \leq 800$ (MEV), where the ρ - resonance appears.

IX. Further extension of the validity domain for fixed t dispersion relations, elastic unitarity.

Let us return to the case of elastic two-body scattering:

For the absorptive part of the scattering amplitude we have obtained two ellipses of analyticity: The large Lehmann ellipse $E_L(s)$ and the unitarity ellipse $E_U(R,s)$ (see page 36). Combining these two, we will obtain an enlargement of the domain $\mathcal{D}$ where fixed t dispersion relations hold. For definiteness let us consider the case of elastic $\pi-\pi$ scattering $(\mu = 1)$. Using crossing symmetry, $\mathcal{D}$ is given by

$$\mathcal{D}^{\pi\pi} = \bigcap_{4 \leq s < \infty} \left(E_L(s) \cup E_U(4,s) \right).$$

Now both ellipses have foci at $0, 4-s$. Therefore the one having a larger right extremity is the bigger one. The right extremity for $E_U(4,s)$ is 4 and for $E_L(s)$ $256 s^{-1}$. So for $s \leq 64$, $E_L(s)$ contains $E_U(4,s)$; for $s \geq 64$ the opposite is true. Also $E_U(4,s) \supset E_U(4,64)$ for $s \geq 64$ and $E_L(s) \supset E_L(16)$ for $4 \leq s \leq 16$ so finally

$$\mathcal{D}^{\pi\pi} = \bigcap_{16 \leq s \leq 64} E_L(s).$$

To find the exact shape of $\mathcal{D}^{\pi\pi}$ is rather tedious. The result is:
The border of $\mathcal{D}^{\pi\pi}$ is made partly of the ellipse $E_L(64)$ for $\mathrm{Re}\, t \geq -20$, partly of the ellipse $E_L(16)$ and a piece of the envelope of the ellipses defined by

$$\cos\left[\arg(t+s-4)\right] = 1 - 2 \cdot \frac{256}{s^2} .$$

In the pion nucleon case we obtain [S 4]

$$\mathcal{D}^{\pi N} = \bigcap_{(M+1)^2 \leq s \leq (M+2)^2 + 8(M+1)} E_L(s) \qquad ,$$

(M = nucleon mass in pion units).

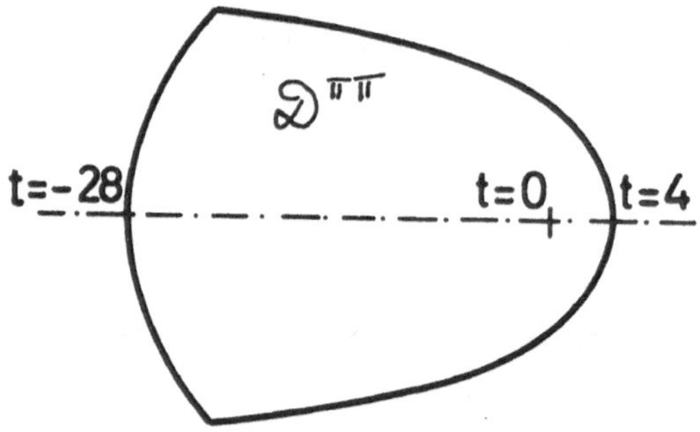

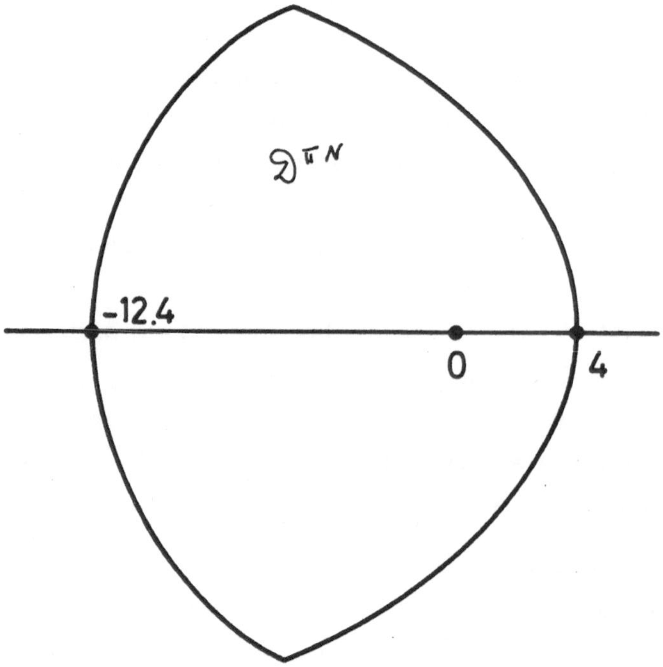

As a new ingredient we want to apply the elastic unitarity condition:

$$|f_\ell(s)|^2 = \mathrm{Im}\, f_\ell(s)$$

to the discussion of the analyticity of the amplitudes in t (resp. $\cos\Theta$).

The reason for doing this is the following: We have seen that the interval where classical fixed t dispersion relations hold, is given by

$$t_0 = \max_{s \geq s_{thr}} \ell_L(s) ,$$

$\ell_L(s)$ being the left extremity of the large Lehmann ellipse corresponding to the given process. We thus obtain

$$t_0(\pi\pi \to \pi\pi) = -28\mu^2,$$
$$t_0(\pi N \to \pi N) = -12.4\mu^2,$$
$$t_0(\pi K \to \pi K) = -31.5\mu^2.$$

Now let t_0' be defined by

$$t_0' = \max_{s \geq s_{inel.\ thr.}} \ell_L(s) .$$

t_0' is called the inelastic point. For example

$$t_0'(\pi\pi \to \pi\pi) = -28\mu^2,$$
$$t_0'(\pi N \to \pi N) = -18\mu^2,$$
$$t_0'(\pi K \to \pi K) = -34.9\mu^2.$$

Thus if it would be possible to prove analyticity of $A_s(s,t)$ (resp. $A_u(u,t)$) in t in a region containing the interval $(t_0', 0]$ for all s $(s_{thr.} \leq s \leq s_{inel.\,thr.})$ we would obtain an extension of the validity of fixed t dispersion relations to the interval $(t_0', 0]$.

Let us first consider the case where the amplitude is analytic in an ellipse with foci ± 1 and semimajor axis $x_0(s)$:

$$F(s, \cos\Theta) = \frac{s^{\frac{1}{2}}}{2k} \sum_{\ell=0}^{\infty} (2\ell+1)\, f_\ell(s)\, P_\ell(\cos\Theta) .$$

Then

$$\overline{\lim_{\ell}} \; |f_\ell(s)|^{\frac{1}{\ell}} \leq \frac{1}{x_0(s) + (x_0^2(s)-1)^{\frac{1}{2}}}$$

and elastic unitarity will give

$$\overline{\lim_{\ell}} \; |\operatorname{Im} f_\ell(s)|^{\frac{1}{\ell}} \leq \frac{1}{\left(2x_0^2(s)-1 + ((2x_0^2(s)-1)^2-1)^{\frac{1}{2}}\right)}$$

such that

$$\operatorname{Im} F(s,\cos\Theta) = \frac{s^{\frac{1}{2}}}{2k} \sum_{\ell=0}^{\infty} (2\ell+1) \operatorname{Im} f_\ell(s) P_\ell(\cos\Theta)$$

will be analytic in $\cos\Theta$ in an ellipse with foci $\pm$ and semimajor axis $2x_0^2(s)-1$. The converse also holds, showing that the ellipse for $\operatorname{Re} F(s,\cos\Theta)$ is at least "half" as large as that of $\operatorname{Im} F(s,\cos\Theta)$. Here we even do not need the condition that s is below the inelastic threshold.

In this discussion, however, we did not take into account the distributional character of $F(s,\cos\Theta)$ for physical s. Instead we would have to consider quantities such as

$$\int_{s_1}^{s_2} \omega(s) f_\ell(s)\, ds \quad \text{and} \quad \int_{s_1}^{s_2} \omega(s) \operatorname{Im} f_\ell(s)\, ds$$

with a smooth (positive) function $\omega(s)$. Even here we may still get information on $F_{av}(\cos\Theta)$ (and $\operatorname{Re} F_{av}(\cos\Theta)$) starting from $\operatorname{Im} F_{av}(\cos\Theta)$. This is because we have the inequality

$$\left|\int_{s_1}^{s_2} \omega(s) f_\ell(s)\, ds\right|^2 \leq \int_{s_1}^{s_2} \omega(s)\, ds \int_{s_1}^{s_2} \omega(s) |f_\ell(s)|^2 ds$$

$$\leq \int_{s_1}^{s_2} \omega(s)\, ds \int_{s_1}^{s_2} \omega(s) \operatorname{Im} f_\ell(s)\, ds .$$

The trouble, however, is that the reverse is not true: given informations on the smoothed $f_\ell(s)$ we cannot get a stronger decrease of $\operatorname{Im} f_\ell(s)$ with ℓ. This is in fact a very deep problem and one can build counter examples in which the connection between $F(s,\cos\Theta_s)$ and $A_s(s,\cos\Theta_s)$ is abnormal. For a discussion of these pathologies see ref [M 11].

However, at the price of a supplementary assumption we can avoid these difficulties. A sufficient condition is:

IX (1) $$\sigma_{el}(s) < B \quad \text{for} \quad s_{thr} \leq s \leq s_{inel.\,thr.}$$

or that

IX (2) $$|f_\ell(s)| < \frac{1}{2} - 2\varepsilon \quad ; \quad \ell > L \quad ; \quad \varepsilon > 0$$

for some sufficiently large L.

Note that the last condition makes sense since we have

$|f_\ell(s)|^2 \leq \mathrm{Im}\, f_\ell(s) \leq 1$ so that $|f_\ell(s)|$ is a measure. Notice also that from condition IX (1) it follows that $\mathrm{Im}\, f_\ell \leq \frac{B k^2}{(2\ell+1)4\pi}$ so that IX (2) is satisfied for ℓ large enough.

It will be sufficient to discuss the situation locally.

From the results of Bros, Epstein and Glaser we know that to each physical point s_0 we may find a neighborhood of s_0 and the interval $[-1, +1]$ of the form:

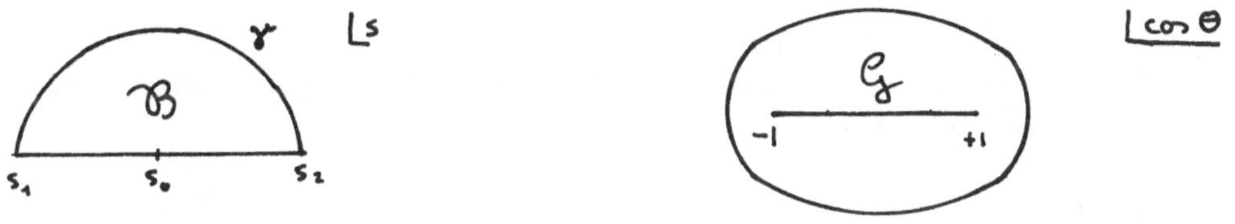

in which $\Psi(s, \cos\Theta)$ is analytic and has distributional boundary values for $\mathrm{Im}\, s \to 0^+$:

IX (3) $$|\Psi(s, \cos\Theta)| < \frac{C}{(\mathrm{Im}\, s)^n} \quad ; \quad s \in \mathfrak{B}\ , \ \cos\Theta \in \mathcal{G}\ .$$

This implies the same behaviour for $f_\ell(s)$ in $\mathfrak{B}$

IX (4) $$|f_\ell(s)| = \frac{1}{2}\left|\int_{-1}^{+1} P_\ell(\cos\Theta)\, \frac{2k}{s^{\frac{1}{2}}}\, \Psi(s, \cos\Theta)\, d\cos\Theta\right| < \frac{C}{(\mathrm{Im}\, s)^n}\ .$$

Now since on the interval $s_1 \leq s \leq s_2$ $|f_\ell(s)| \leq 1$, the function

$$\phi_\ell(s) = (s - s_1)^n (s - s_2)^n f_\ell(s)$$

is bounded on $\partial\mathfrak{B} = \gamma \cup [s_1, s_2]$. Therefore $\phi_\ell(s)$ is bounded in $\mathfrak{B}$

and in particular $f_\ell(s)$ is bounded in $\mathcal{B}'$ where $\mathcal{B}'$ is a half-disc with center at s_0 properly contained in $\mathcal{B}$:

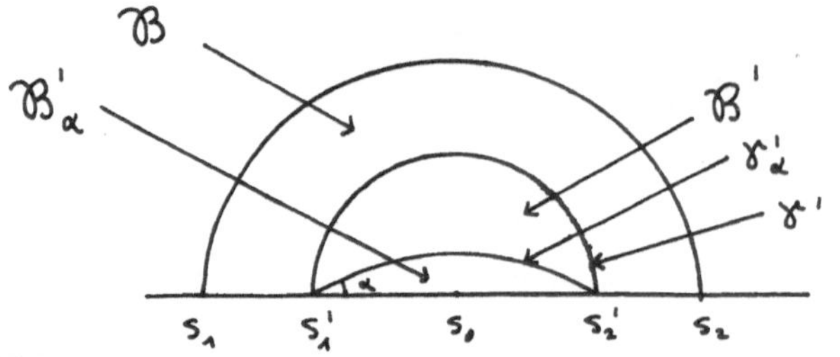

IX (5) $\quad |f_\ell(s)| < C' \qquad$ in $\mathcal{B}'$ $\quad$ (C' independent of ℓ).

If $\omega(s)$ is C^∞ and positive with sufficiently small support about the origin, we also have

IX (6) $$\left| \int \omega(s') f_\ell(s-s')\, ds' \right| < C \int \omega(s')\, ds' \qquad ; \ s \in \mathcal{B}'.$$

On the other hand we have

IX (7) $$\left| \int \omega(s') f_\ell(s-s')\, ds' \right| < C(\omega)\, x^\ell \ ; \quad s \in [s_1', s_2']$$

where $x < 1$ is determined by the size of the small Lehmann ellipse. Combining the two bounds IX (6) and IX (7) we have in $\mathcal{B}_\alpha'$

IX (8) $$\left| \int \omega(s') f_\ell(s-s')\, ds' \right| < C(\alpha, \omega) \quad x^{\ell \frac{(\frac{\pi}{2} - \alpha)}{\frac{\pi}{2}}} .$$

This may easily be seen using the comparison function

$$g(s) = \tau \ \exp\, i\lambda \log \frac{s_2' - s}{s - s_1'}$$

adjusted in such a way that

$$|g(s)| = C(\omega)\, x^\ell \qquad \text{on } \gamma'$$

$$|g(s)| = C \int \omega(s')\, ds' \qquad \text{on } [s_1', s_2']$$

Then

$$\left| g(s) \int \omega(s') f_\ell(s-s')\, ds' \right| < C \cdot C(\omega) \cdot \int \omega(s')\, ds' \cdot x^\ell$$

on $\gamma' \cup [s_1', s_2']$ and therefore also in $\mathcal{B}'$.

Dividing $|g(s)|$ out again, we obtain IX (8).

The angle α is determined as follows:

We combine IX (2) and IX (5) using the same argument and choose α in such a way that

IX (9) $$|f_\ell(s)| < \frac{1}{2} - \varepsilon \quad \text{in} \quad \mathcal{B}'_\alpha, \ \ell > L.$$

Then IX (8) may be deregularized, giving

IX (10) $$|f_\ell(s)| < \hat{C} \cdot \frac{x'^{\ell}}{(\operatorname{Im} s)^p}; \ s \in \mathcal{B}'_\alpha; \ x' = x^{(1-\frac{2\alpha}{\pi})}.$$

As a last step we make an analytic continuation to $\mathcal{B}'^*_\alpha = \{s \mid s^* \in \mathcal{B}'_\alpha\}$ of $f_\ell(s)$ $(\ell > L)$ [Z 1] :

IX (11) $$f_\ell^{II}(s) = \frac{f_\ell^*(s^*)}{1 - 2i\, f_\ell^*(s^*)}; \quad s \in \mathcal{B}'^*_\alpha .$$

Because of IX (9) this is possible. Note that the assumption IX (2) was just made to get this analytic continuation. Unitarity then implies that $f_\ell(s)$ and $f_\ell^{II}(s)$ coincide on $\mathcal{B}'_\alpha \cap \mathcal{B}'^*_\alpha = [s_1', s_2']$. Now $f_\ell^{II}(s)$ is bounded on $\gamma'^*_\alpha = \{s \mid s^* \in \gamma'_\alpha\}$:

IX (12) $$|f_\ell^{II}(s)| < \hat{C} \frac{x'^{\ell}}{|\operatorname{Im} s|^p} \frac{1}{\varepsilon} .$$

Then $\Phi_\ell'(s) = (s-s_1')^p (s-s_2')^p f_\ell(s)$ is bounded on $\gamma'_\alpha \cup \gamma'^*_\alpha$. Since Φ_ℓ' is analytic in $\mathcal{B}'_\alpha \cup \mathcal{B}'^*_\alpha$ the same bound holds in $\mathcal{B}'_\alpha \cup \mathcal{B}'^*_\alpha$. Shrinking this region, we obtain from IX (10) and IX (12) a bound for $f_\ell(s)$ on a real interval containing s_0. This bound is of the form

IX (13) $$|f_\ell(s)| < \tilde{C} \cdot x'^{\ell}$$

where x' $(> x)$ may be chosen arbitrarily close to x. But this estimate shows that we may return to our previous arguments which ignored the distributional character of $F(s, \cos\Theta)$.

If the region $\mathcal{D}$ of fixed t dispersion relations is not an ellipse, we may apply Hadamard's multiplication theorem [T 2] : The set of singular points of $A_s(s, \cos\Theta)$ in $\cos\Theta$ is contained in

$$\left\{ \cos\Theta \ \Big| \ \cos\Theta = \cos(\Theta_1 + \Theta_2); \ \cos\Theta_1, \ \cos\Theta_2^* \in \mathcal{D} \right\}.$$

For the proof consider the two functions [J 2] :

$$F(s, \cos\Theta) = \frac{s^{\frac{1}{2}}}{2k} \sum_{\ell=0}^{\infty} (2\ell+1) \, f_\ell(s) \, P_\ell(\cos\Theta)$$

$$\hat{F}(s, z) = \frac{s^{\frac{1}{2}}}{2k} \sum_{\ell=0}^{\infty} (2\ell+1) \, f_\ell(s) \, z^\ell.$$

We have the correspondence

$$F(s, \cos\Theta) = \frac{1}{\pi} \int_0^{\pi} \hat{F}(s, \cos\Theta + i \sin\Theta \cos\phi) \, d\phi$$

$$\hat{F}(s, z) = \frac{1}{2} \int_{-1}^{+1} \frac{(1 - z^2)}{(1 - 2z\cos\Theta + z^2)^{\frac{3}{2}}} \, F(s, \cos\Theta) \, d\cos\Theta$$

which may be proved using the orthogonality relations of the Legendre polynomials and the relation

$$\frac{(1 - z^2)}{(1 - 2\cos\Theta \, z + z^2)^{\frac{3}{2}}} = \sum_{\ell=0}^{\infty} (2\ell+1) \, P_\ell(\cos\Theta) \, z^\ell.$$

The correspondence between the singularities of $F(s, \cos\Theta)$ and $\hat{F}(s, z)$ is given by the relation

$$\text{IX (14)} \qquad z = \cos\Theta + (\cos^2\Theta - 1)^{\frac{1}{2}} = e^{i\Theta}.$$

Let $\hat{\mathcal{D}}$ be the domain of analyticity in z of $\hat{F}(s, z)$ which corresponds to the domain $\mathcal{D}$ of $F(s, \cos\Theta)$. Note that so far we have required s to be off the cut. The statement we want to prove for the absorptive part is, however, for s on the cut. In order to achieve this consider

$$\hat{F}_L(s, z) = \hat{F}(s, z) - \frac{s^{\frac{1}{2}}}{2k} \sum_{\ell=0}^{L} (2\ell+1) \, f_\ell(s) \, z^\ell$$

near $s = s_0$ ($s_{thr} < s_0 < s_{inel. thr.}$).

Due to IX (2) we were able to define $f_\ell(s)$ $(\ell > L)$ to be analytic in $\mathcal{B}'_\alpha \cup \mathcal{B}'^*_\alpha$. But then the bounds IX (13) , which also hold in a neighborhood of

$s = s_0$, and Hartog's theorem show that $\hat{\Psi}_L(s, z)$ is analytic in both variables in a neighborhood of $s = s_0$, $z = 0$. If we now choose a path in $\mathcal{B}'_\alpha$ with one end point $s = s_0$, Bremermann's continuity theorem [see chap. XI] shows that at the point $s = s_0$ $\hat{\Psi}_L(s, z)$, and hence $\hat{\Psi}(s, z)$, is analytic in z for $z \in \hat{\mathcal{D}}$.

Starting now from $\hat{\Psi}(s_0, z)$ and

$$\hat{\Psi}^*(s_0, z^*) = \frac{s_0^{\frac{1}{2}}}{2k_0} \sum_{\ell=0}^{\infty} (2\ell+1) f_\ell^*(s_0) z^\ell ,$$

Hadamard's theorem gives us the singularities of

$$\hat{B}(s_0, z) = \frac{s_0^{\frac{1}{2}}}{2k_0} \sum_{\ell=0}^{\infty} (2\ell+1)^2 |f_\ell(s_0)|^2 z^\ell .$$

But the singularities of

$$\text{IX (15)} \qquad \hat{A}_s(s_0, z) = \frac{s_0^{\frac{1}{2}}}{2k_0} \sum_{\ell=0}^{\infty} (2\ell+1) |f_\ell(s_0)|^2 z^\ell$$

are then the same, as is easily seen by integration. Because of IX (14) we then also know the singularities of

$$A_s(s_0, \cos\Theta) = \frac{s_0^{\frac{1}{2}}}{2k_0} \sum_{\ell=0}^{\infty} (2\ell+1) |f_\ell(s_0)|^2 P_\ell(\cos\Theta)$$

in $\cos\Theta$. Since s_0 was arbitrary in $(s_{thr}, s_{inel.\,thr.})$ the proof is complete.

This way of using elastic unitarity indeed made it possible to extend the validity of fixed t dispersion relations to the inelastic point for the processes [S 5] [S 6] :

$$\pi N \to \pi N \quad ; \quad \pi K \to \pi K .$$

Another interesting application has been given by Cheung and Toll [C 2]. They show the necessity of production amplitudes in the following way: The assumption that elastic unitarity should hold for all s leads to the conclusion $\Psi(s, \cos\Theta) \equiv 0$.

We assume that $\sigma_{tot}(s) < s^N$ for some N.

For the proof let us start from the unitarity ellipse $E_U(R, s)$. We have

IX (16) $$\bigcap_{s \geq s'} E_U(R, s) = E_U(R, s') .$$

Now let us assume we have fixed t dispersion relations in $E_U(R, s_0)$. This is in particular true for $s_0 = s_{thr}$, where $E_U(R, s_0)$ is a circle with radius R and center at the origin. Let us choose $s > s_0$ such that

$$r = 4k_0^2 + R - 4k^2 > 0 . \text{ Then } E_U(r, s) \supset E_U(R, s_0).$$

Applying elastic unitarity, we find that the absorptive part $A_s(s, t)$ is analytic in $E(0, -4k^2 | 4r(1 + \frac{r}{4k^2}))$. Here $E(a, b | c)$ denotes an ellipse with right focus a, left focus b and right extremity c.

We have therefore obtained an enlargement compared to $E_U(R, s_0)$ if $4r(1 + \frac{r}{4k^2}) > R$ i.e. if $k_0^2 < k^2 < k_0^2 + \frac{3}{16} R$.

Because of IX (16) we have extended the validity of fixed t dispersion relations to the domain $\mathcal{E}_U(R, s)$. Here $\mathcal{E}_U(R, s)$ is defined to be the ellipse $E_U(R, s)$ minus a left hand cut which arises from the contribution of the crossed channel in the dispersion relations. Using Hadamard's multiplication theorem, we may repeat this enlargement procedure, however, the resulting "ellipses" $\mathcal{E}_U$ always have right extremity at $t = R$. If we apply elastic unitarity once more, we may go from $\mathcal{E}_U(R, s)$ to $\mathcal{E}_U(0, -4k^2 | 4R(1 + \frac{R}{4k^2})) \supset \mathcal{E}_U(4R, s)$ Note that $\mathcal{E}_U(4R, s)$ also may have a right hand cut due to Hadamard's multiplication theorem. Repeating the above arguments with $\mathcal{E}_U(4R, s)$ instead of $\mathcal{E}_U(R, s)$ we immediately see that we arrive at fixed t dispersion relations for all t in the cut plane with a left and right hand cut. In case there are pole terms, the situation is essentially the same. In particular we may apply this discussion to potential scattering, if there exists a fixed dispersion relation to start with.

Thus the assumption of elastic unitarity for all s gives the Mandelstam representation. The arguments further needed to obtain contradiction may now be taken from an article by Aks [A 1]. We will, however, be satisfied with the following argument: Consider pion-nucleon scattering. Then in particular for $4\mu^2 < t < 16\mu^2$ the above arguments show that the amplitude is real since the right hand cut appears for larger values of t. But t is the energy variable in the $\pi\pi \to N\bar{N}$ channel and there an unphysical threshold appears at $t = 4\mu^2$. In the interval $4\mu^2 < t < 16\mu^2$ in particular, the absorptive part is determined by the 2π - intermediate states. But then the vanishing of the absorptive part implies the vanishing of all 2π - elastic partial-wave phase shifts [M 4], which is a contradiction.

Finally we may quote here another result, which makes a much stronger statement in a particular case:

If the 2π - inelastic cross-section vanishes for some $s > 20\mu^2$, then the whole amplitude vanishes [D 2] [M 8].

X. Extension of the analyticity domain for the absorptive part of the two pion scattering amplitude by means of unitarity and positivity

We start with the method of the last chapter for the elastic region. First we apply crossing symmetry to combine the domains $\mathcal{D}^{\pi\pi}(t)$ for fixed t dispersion relations and $\mathcal{D}^{\pi\pi}(u)$ for fixed u dispersion relations (see page 48) to get a domain of the following shape

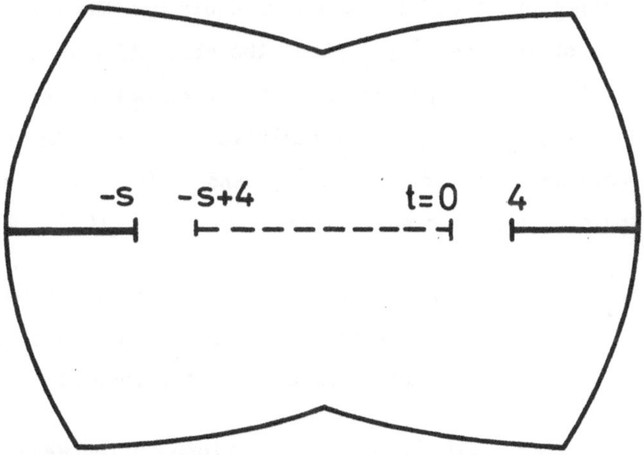

Now this domain contains the ellipse $E(0, 4-s \mid 4)$. Applying the unitarity condition, we obtain the ellipse $E(0, 4-s \mid 16 + \frac{64}{s-4})$ whose extremities $t = 16 + \frac{64}{s-4}$ and $t = 4 - s - 16 - \frac{64}{s-4}$ are just the equations for the boundary of the Mandelstam spectral function in $4 < s < 16$.

Using Hadamard's multiplication theorem we obtain an analyticity domain for the absorptive part of the form

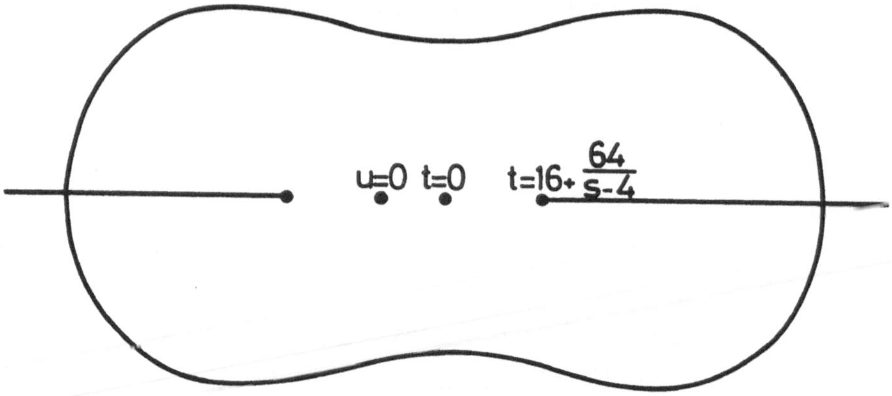

In addition for $16 < s < 64$ we have the large Lehmann ellipses $E_L(s)$ with extremities $256\,s^{-1}$ and $4 - s - 256\,s^{-1}$, and for $s > 64$ we have the unitarity ellipses $E_U(4, s)$ with extremities 4 and $-s$.

Now, if the Mandelstam representation were true, we could use the fact that if around some point (s_0, t_0) of the real (s,t) plane the absorptive part $A_s(s, t)$ is analytic in t, then the double spectral function vanishes and hence the absorptive part in the t channel $A_t(s,t)$ is analytic in s in the same region. The existing information concerns $4 < s < 16$ and $4 < t < \infty$. Then by crossing symmetry we could enlarge the analyticity domain of $A_s(s,t)$ for $s > 16$.

What we want to do here is to find a substitute to prove a similar result without knowing in advance the validity of Mandelstam representation. The main tool is again the positivity properties of $A_s(s,t)$ and its derivatives with respect to t.

To exploit crossing symmetry we write dispersion relations in two ways, for s and t inside the triangle $s < 4$, $t < 4$, $u < 4$ restricting ourselves for the moment to the $\pi^0\pi^0 \to \pi^0\pi^0$ scattering amplitude:

$$\text{x (1)} \qquad F(s,t,u) = \frac{1}{\pi}\int_4^\infty \frac{A(x,t)}{x-s}\,dx + \frac{1}{\pi}\int_4^\infty \frac{A(x,t)}{x-u}\,dx$$

$$= \frac{1}{\pi}\int_4^\infty \frac{A(x,s)}{x-t}\,dx + \frac{1}{\pi}\int_4^\infty \frac{A(x,s)}{x-u}\,dx$$

where $A(x,t)$ denotes the absorptive part of the amplitude in that channel where x is the energy.

Note:

X (2) $$A(x,t) = A(x, 4-x-t).$$

Also we have omitted subtractions, but this is legitimate, since we are going to consider derivatives. We apply the differential operator $\left(\frac{d}{ds}\right)^n \left(\frac{d}{dt}\right)^n$ to X (1), and since we may interchange differentiation and integration (see chapter V), we obtain

X (3) $$\frac{1}{\pi} \int_4^\infty \frac{\left(\frac{d}{ds}\right)^n A(x,s)}{(x-t)^{n+1}} dx$$

$$= \frac{1}{\pi} \int_4^\infty \frac{\left(\frac{d}{dt}\right)^n A(x,t)}{(x-s)^{n+1}} dx + \frac{1}{n!} \left(\frac{d}{ds}\right)^n \left(\frac{d}{dt}\right)^n \frac{1}{\pi} \int_4^\infty \frac{A(x,t)-A(x,s)}{x-u} dx.$$

We claim the following to be true :

X (4) $$(-1)^{n+1} \left(\frac{d}{dt}\right)^n \left(\frac{d}{ds}\right)^n \int_4^\infty \frac{A(x,t)-A(x,s)}{x-u} dx \geq 0$$

for $t = 0 \; , \quad 0 < s < 4.$

For the proof we note that in the Legendre expansion of $A(x,t)$ the coefficients are nonnegative due to the positivity property and vanish for ℓ odd because of crossing symmetry, so all we have to prove is

X (5) $$(-1)^{n+1} \left(\frac{d}{ds}\right)^n \left(\frac{d}{dt}\right)^n \left. \frac{P_\ell\left(1+\frac{2t}{x-4}\right) - P_\ell\left(1+\frac{2s}{x-4}\right)}{x-4+s+t} \right|_{t=0} \geq 0$$

ℓ even, $0 < s < 4.$

Since $P_\ell(-y) = P_\ell(y)$ for ℓ even and since all zero's of $P_\ell(y)$ lie in the interval $[-1,1]$ we may write (x_i being the positive zeros of P_ℓ) :

$$P_\ell(y) = \prod_{i=1}^{\frac{\ell}{2}} (y^2 - x_i^2) = \prod_{i=1}^{\frac{\ell}{2}} (y^2 - 1 + 1 - x_i^2)$$

$$= \sum_{p=0}^{\frac{\ell}{2}} c_p \, (y^2-1)^p \quad ; \; c_p \geq 0.$$

Furthermore

$$\frac{\left(\left(1+\frac{2t}{x-4}\right)^2-1\right)^p - \left(\left(1+\frac{2s}{x-4}\right)^2-1\right)^p}{x-4+s+t}$$

$$= \frac{4^p}{(x-4)^{2p}} \; \frac{(x-4+t)^p \cdot t^p - (x-4+s)^p \cdot s^p}{x-4+s+t} .$$

So it suffices to prove

$$(-1)^{n+1} \left(\frac{d}{dt}\right)^n \left(\frac{d}{ds}\right)^n \left. \frac{(x-4+t)^p \cdot t^p - (x-4+s)^p \cdot s^p}{x-4+s+t} \right|_{t=0} \geq 0 \; ; \; 0 < s < 4.$$

We distinguish two cases:

1°) $p \leq n$,

2°) $p > n$.

In the first case

$$\frac{(x-4+t)^p \cdot t^p - (x-4+s)^p \cdot s^p}{x-4+s+t} = \sum_{\alpha=0}^{p-1} (t-s) \{(x-4+t)\cdot t\}^\alpha \{(x-4+s)\cdot s\}^{p-1-\alpha} .$$

But

$$\left(\frac{d}{dt}\right)^n \left(\frac{d}{ds}\right)^n (t-s)\{(x-4+t)\cdot t\}^\alpha \{(x-4+s)\cdot s\}^{p-1-\alpha} \equiv 0$$

for $0 \leq \alpha \leq p-1$, $p \leq n$ as is easily verified looking at the degree of each monomial.

In the second case

$$\left(\frac{d}{dt}\right)^n \frac{(x-4+t)^p \cdot t^p}{(x-4+s+t)}\Bigg|_{t=0} \equiv 0$$

and

$$(-1)^{n+1}\left(\frac{d}{ds}\right)^n\left(\frac{d}{dt}\right)^n\left\{-\frac{(x-4+s)^p \cdot s^p}{x-4+s+t}\right\}\Bigg|_{t=0} = n!\left(\frac{d}{ds}\right)^n (x-4+s)^{p-n-1} s^p > 0$$

for $0 < s < 4 \; ; \; x > 4.$

This proves X (5) and thus X (4).

Combined with X (3) we obtain

$$\text{X (6)} \qquad \int_4^\infty \frac{\left(\frac{d}{ds}\right)^n A(x,s)}{x^{n+1}}\,dx < \int_4^\infty \frac{\left(\frac{d}{dt}\right)^n A(x,t=0)}{(x-s)^{n+1}}\,dx \; ; \quad 0 < s < 4$$

for n even. For n odd the inequality is reversed.

The right hand side X (6) may be bounded by the Cauchy inequalities. Thus if $A(x,t)$ is analytic in $|t| \leq R(x)$ the positivity properties of $A(x,t)$ imply

$$0 \leq \left(\frac{d}{dt}\right)^n A(x,t=0) < \frac{n!}{(R(x))^n} A(x,R(x))$$

If we choose

$$R(x) = 256\,x^{-1} \qquad \text{for} \quad x > 16$$

$$R(x) = 16 + \frac{64}{x-4} \qquad \text{for} \quad 16 > x > 4$$

then for even n we obtain

$$\frac{1}{\pi}\int_4^\infty \frac{\left(\frac{d}{ds}\right)^n A(x,s)}{x^{n+1}}\,dx \leq n!\,\frac{1}{\pi}\int_4^{16} \frac{A\left(x, 16+\frac{64}{x-4}\right)dx}{\left\{\left(16+\frac{64}{x-4}\right)(x-s)\right\}^n (x-s)}$$

$$+\; n!\,\frac{1}{\pi}\int_{16}^\infty \frac{A(x, 256x^{-1})\,dx}{\{256x^{-1}(x-s)\}^n (x-s)}$$

For

$$x \geq 16 \qquad 256\, x^{-1}(x-s) \geq 256\left(1-\frac{s}{16}\right)$$

$$16 > x > 4 \qquad \left(16+\frac{64}{x-4}\right)(x-s) \geq 16\left(2+\sqrt{4-s}\right)^2$$

Since $16\left(2+\sqrt{4-s}\right)^2 \leq 256\left(1-\frac{s}{16}\right)$ we finally obtain for even n and $0 \leq s < 4$ the result:

$$\text{X (7)} \quad \frac{1}{\pi}\int_4^\infty \frac{\left(\frac{d}{ds}\right)^n A(x,s)}{x^{n+1}}\,dx \leq \frac{n!\ C}{\left\{16\left(2+\sqrt{4-s}\right)^2\right\}^n}$$

where C is independent of n. From X (7) we conclude that for n even

$$\int_{x_1}^{x_2} \Phi(x)\left(\frac{d}{ds}\right)^n A(x,s)\,dx \leq \frac{C\cdot \sup\Phi\cdot n!\ x_2^{n+1}}{\left\{16\left(2+\sqrt{4-s}\right)^2\right\}^n}$$

where $\Phi(x)$ is a continuous positive function with support in $[x_1,x_2]$.

Hence if we define

$$A(\Phi,s) = \int_{x_1}^{x_2} \Phi(x)\, A(x,s)\,dx$$

we have for even n and $0 \leq s_0 < 4$

$$\left(\frac{d}{ds}\right)^n A(\Phi,s_0) \leq \frac{C\cdot\sup\Phi\cdot n!\ x_2^{n+1}}{\left\{16\left(2+\sqrt{4-s_0}\right)^2\right\}^n}$$

The odd derivatives are majorized as follows:

$$\left(\frac{d}{ds}\right)^{n-1} A(\Phi,s_0) \leq s_0\left(\frac{d}{ds}\right)^n A(\Phi,s_0) + \left(\frac{d}{ds}\right)^{n-1} A(\Phi,0)$$

since for $0 \leq s' \leq s_0$ $\left(\frac{d}{ds}\right)^n A(\Phi,s')$ is a positive increasing function. In addition, since we always have analyticity in $|s| < 256\,x^{-1}$, Cauchy's inequalities and positivity give

$$\left(\frac{d}{ds}\right)^{n-1} A(x,0) \leq \frac{(n-1)!\, A(x, \frac{256}{x})}{(256\, x^{-1})^{n-1}}$$

and hence

$$\left(\frac{d}{ds}\right)^{n-1} A(\phi, s_0) < x_2^{n-1} (n-1)! \left\{ \frac{C \cdot \sup \phi \cdot n \cdot s_0 \cdot x_2^2}{\{16[2+\sqrt{4-s_0}]^2\}^2} + \frac{\int_{x_1}^{x_2} \phi(x)\, A(x, 256x^{-1})\, dx}{256^{n-1}} \right\}$$

Since $256 > 16\,[2+\sqrt{4-s_0}]^2$, and since $A(x, 256x^{-1})$ is a measure, we obtain for n even or odd, the result

$$\left(\frac{d}{ds}\right)^{n} A(\phi, s_0) < \frac{C(x_1,x_2) \cdot \sup \phi \cdot x_2^n \cdot (n+1)!}{\{16\,[2+\sqrt{4-s_0}]^2\}^n} .$$

Hence $A(x,s)$ is a measure in x analytic in s in the domain

$$|s-s_0| < \frac{16\,[2+\sqrt{4-s_0}]^2}{x} .$$

Crossing then implies that $A_s(s,t)$ is analytic in t in

$$|t-t_0| < \frac{16\,[2+\sqrt{4-t_0}]^2}{s}$$

for $0 \leq t_0 < 4$.

The positivity properties of $A_s(s,t)$ now allow an even stronger statement:

$A_s(s,t)$ is analytic in

$$|t| < t_0 + \frac{16}{s}\,(2+\sqrt{4-t_0})^2 .$$

for any t_0 in $[0,4)$.

The optimum for t_0 turns out to be $t_0 = 0$ for $s < 32$, so we obtain $|t| < 256\,s^{-1}$ which is just the result from the large Lehmann ellipse, so nothing has been gained here. For $s > 32$, however, we obtain an improvement:

$$t_0 = 4 - \left(\frac{32}{s-16}\right)^2$$

which gives

X (8) $$|t| < 4 + \frac{64}{s-16}$$

i.e. exactly the Mandelstam equation for the boundary of the spectral function. The region X (8) can, of course, be replaced by $E\left(0, 4-s \mid 4 + \frac{64}{s-16}\right)$.

So finally the right extremities of the ellipses of analyticity of $A_s(s,t)$ have the values

X (9)
$$r(s) = 16 + \frac{64}{s-4} \quad ; \quad 4 < s < 16$$
$$r(s) = 256\, s^{-1} \quad ; \quad 16 < s < 32$$
$$r(s) = 4 + \frac{64}{s-16} \quad ; \quad 32 < s .$$

Graphically we have the following picture in the (t,s) - plane

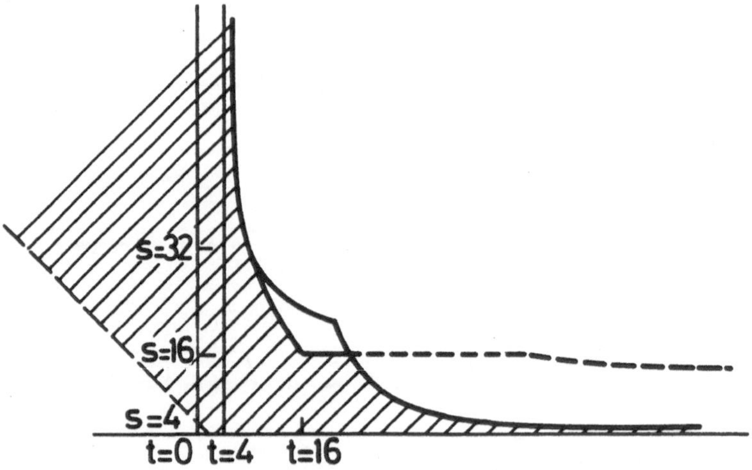

The solid line describes the border line of the Mandelstam spectral function.

So far this proof is only valid for the $\pi^0\pi^0 \to \pi^0\pi^0$ amplitude. However, the relations (I = isospin)

$$\operatorname{Im} f_e^{\pi^0\pi^0 \to \pi^0\pi^0} = \frac{1}{3} \operatorname{Im} f_e^{I=0} + \frac{2}{3} \operatorname{Im} f_e^{I=2}$$

give

$$0 \leq \operatorname{Im} f_e^{I=0} \leq 3 \operatorname{Im} f_e^{\pi^0\pi^0 \to \pi^0\pi^0}$$

$$0 \leq \operatorname{Im} f_e^{I=2} \leq \frac{3}{2} \operatorname{Im} f_e^{\pi^0\pi^0 \to \pi^0\pi^0}$$

This shows that the result obtained for $\pi^0\pi^0 \to \pi^0\pi^0$ also holds for $\pi^+\pi^+ \to \pi^+\pi^+$, $\pi^-\pi^- \to \pi^-\pi^-$, $\pi^-\pi^+ \longleftrightarrow \pi^0\pi^0$. To get the analyticity domain for the case $\pi^+\pi^- \to \pi^+\pi^-$ which contains a $I = 1$ component, it is reasonable to start from

$$\frac{1}{\pi}\int_4^\infty \frac{A^{\pi^+\pi^- \to \pi^+\pi^-}(x,t)}{(x-s)}\,dx + \frac{1}{\pi}\int_4^\infty \frac{A^{\pi^+\pi^+ \to \pi^+\pi^+}(x,t) - A^{\pi^+\pi^+ \to \pi^+\pi^+}(x,s)}{(x-u)}\,dx$$

$$= \frac{1}{\pi}\int_4^\infty \frac{A^{\pi^+\pi^- \to \pi^+\pi^-}(x,s)}{(x-t)}\,dx$$

and carry through the whole argument. This is possible because of the symmetries in the integral over the u - cut.

An immediate consequence of this enlargement is that the region of validity of the fixed t dispersion relation can be extended from

$$\mathcal{D}^{\pi\pi} = \bigcap_{4 \leq s < \infty} \left(E_L(s) \cup E_U(4,s)\right)$$ (see page 48)

to

$$\mathcal{D}'^{\pi\pi} = \bigcap_{4<s<16} E(0,4-s\,|\,16) \bigcap_{16<s<32} E_L(s) \bigcap_{32<s} E\left(0,4-s\,\middle|\,4+\frac{64}{s-16}\right).$$

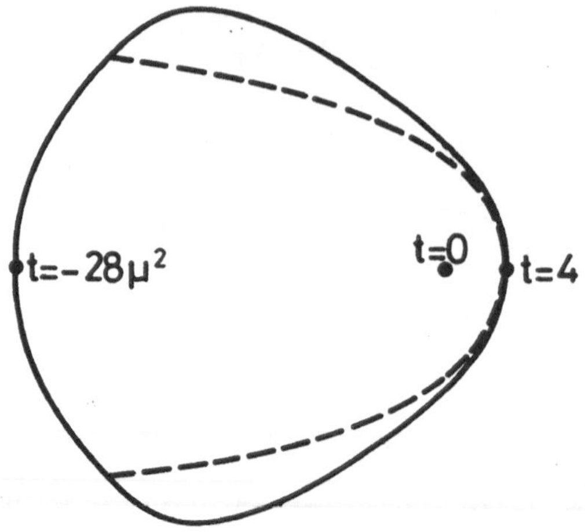

The solid line gives the improvement.

XI Methods of analytic completion

In the following discussion we will use a fact which is peculiar for analytic functions of several complex variables:

Let D be an open set in $\mathbb{C}^n$ $(n \geq 1)$ and let $A(D)$ be the ring of functions analytic in D. D is said to be a domain of holomorphy if there are no open sets D_1 and D_2 in $\mathbb{C}^n$ with the following properties:

(a) $\emptyset \neq D_1 \subset D_2 \cap D$.

(b) D_2 is connected and not contained in D.

(c) for every $u \in A(D)$ there exists a function $u_2 \in A(D_2)$ (necessarily uniquely determined), such that $u = u_2$ in D_1.

Roughly speaking, this definition means that there is no part of the boundary across which every element in $A(D)$ can be continued analytically.

Now in $\mathbb{C}^1$ every simply connected domain D having more than one boundary point is a domain of holomorphy, since it can be mapped conformally onto the unit disk (Riemann Mapping Theorem) and since $f(z) = \sum_{n=0}^{\infty} z^{n!}$ is an analytic function in $|z| < 1$ whose singularities are dense on $|z| = 1$.

In $\mathbb{C}^n$ $(n \geq 2)$ this is no longer true. However, if we admit complex manifolds spread over $\mathbb{C}^n$, there exists a (unique) largest set $\mathcal{H}(D)$ to which all functions in $A(D)$ can be continued analytically. $\mathcal{H}(D)$ is then a domain of holomorphy which is called the envelope of holomorphy.

The determination of $\mathcal{H}(D)$, given D, is far from being trivial, but there exists a theorem which is useful at least to enlarge the domain of analyticity and which we shall use:

We define an open set $D \subset \mathbb{C}^n$ to be a tube, if there exists an open set $d \subset \mathbb{R}^n$ called the base of D, such that $D = \{ z \mid \mathrm{Re}\, z \in d \}$. It is obvious that the convex hull $ch\, D$ of D is a tube with base $ch\, d$. Then we have the

<u>Tube Theorem</u> [H 2] : If D is a connected tube, then every $u \in A(D)$ can be extended to a function in $A(ch\, D)$. A tube is a domain of holomorphy if and only if every component is convex.

Although this theorem appears rather particular in nature, it is very useful because many domains which do not look like tubes can be shown to be tubes after a convenient change of variables, and often we can insert a tube inside a domain which is not a tube.

As an example consider a function $f(z_1, z_2)$ analytic in $|z_1| < a$, $|z_2| < b$ and in $|z_1| < b$, $|z_2| < a$. If we introduce the variables $y_i = \ln z_i$ it is easily seen that $f(z_1, z_2)$ may be analytically continued to

$$|z_1 z_2| < ab \; ; \quad |z_1|, |z_2| < \mathrm{Max}\,(a,b).$$

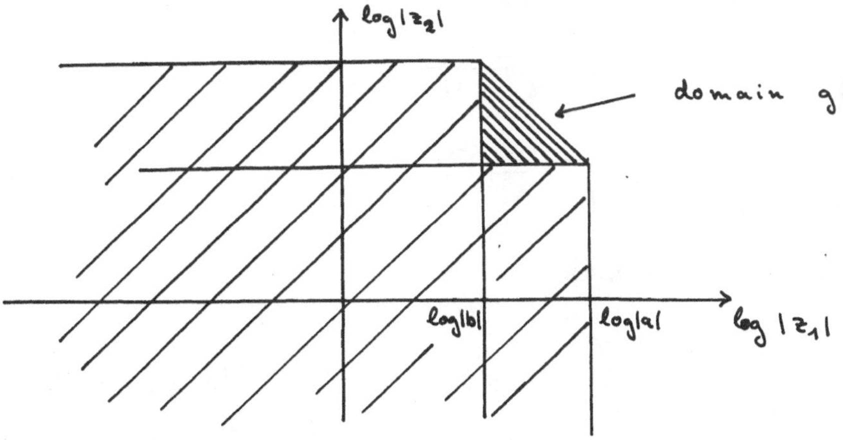

Notice that the multivaluedness of log z_i as a function of z_i plays no role in these considerations, since we only need to specify a connected domain in the $\mathrm{Re}\, y_1 - \mathrm{R}\, y_2$ - plane to get a tube.

Another useful example is a domain given as follows:

Take as a domain the union of $\mathcal{D}_1$ and $\mathcal{D}_2$ with

L_{z_1} L_{z_2}

$\mathcal{D}_1$:

i.e. : $0 \leq \arg \frac{B - z_1}{z_1 - A} \leq \alpha$; $\quad 0 \leq \arg \frac{D - z_2}{z_2 - C} \leq \beta$

$\mathcal{D}_2$:

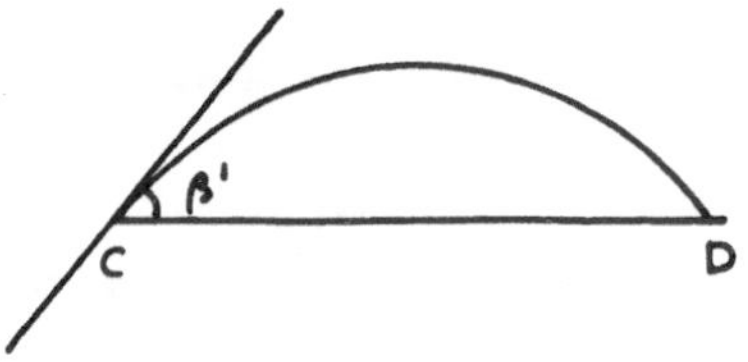

i.e. : $0 \leq \arg \frac{B - z_1}{z_1 - A} \leq \alpha'$; $\quad 0 \leq \arg \frac{D - z_2}{z_2 - C} \leq \beta'$.

The proper variables to be used in this case are

$$y_1 = -i \log \frac{B - z_1}{z_1 - A}$$

$$y_2 = -i \log \frac{D - z_2}{z_2 - C}$$

Then $\mathcal{D}_1$ and $\mathcal{D}_2$ are given by

$$\mathcal{D}_1 : \quad 0 \leq \mathrm{Re}\, y_1 \leq \alpha ; \; 0 \leq \mathrm{Re}\, y_2 \leq \beta$$

$$\mathcal{D}_2 : \quad 0 \leq \mathrm{Re}\, y_1 \leq \alpha' ; \; 0 \leq \mathrm{Re}\, y_2 \leq \beta' .$$

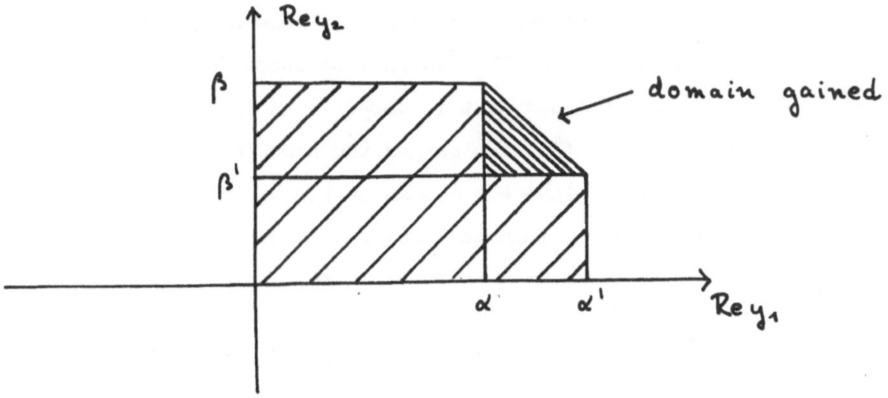

In chapter IX we used a theorem which is known as the

Strong Continuity theorem of Bremermann [B 8] [V 1]

Let $t \to \omega(t)$ be a continuous complex valued function for $0 \leq t \leq 1$, and let a and b belong to $\mathbb{C}^{n-1}$. Define $z(t) = a + \omega(t)b$. Let $D(t)$ $(0 \leq t \leq 1)$ denote a domain in $\mathbb{C}^1$ depending continuously on t near $t = 0$ in the following sense: For every compact $K \subset D(0)$ there is a $\eta > 0$ such that $K \subset D(t)$ for all t such that $0 \leq t \leq \eta$. Let $\Delta(t)$ denote the "disc" in $\mathbb{C}^n$ given by

$$\Delta(t) = \{ z \in \mathbb{C}^n \mid z = (z_1, \cdots z_{n-1}, z_n) = (z^{n-1}, z_n), \\ z^{n-1} = z(t),\ z_n \in D(t) \}$$

If a domain of holomorphy D contains $\Delta(t)$ for all t such that $0 < t \leq 1$ and if D contains one point of $\Delta(0)$, D contains $\Delta(0)$.

We will also use a special theorem which was originally obtained in a particular case by Mandelstam [M 3] and later generalized by Epstein and Glaser [unpublished]:

Let us start with the particular case, by oversimplifying what Mandelstam did, and suppose that for some values of t the analyticity domain of $\Psi(s,t)$ in s is the cut plane with a cut starting at $s = 4\mu^2$. In Mandelstam's paper t was allowed to vary in the interval $[-28\mu^2, 0]$. Then

$$\Psi(s,t) = \frac{1}{\pi} \int_4^\infty \frac{A_s(s',t)}{s'-s}\, ds' + R(s,t)$$

where $R(s,t)$ represents other contributions. We have another information, namely the discontinuity $A_s(s',t)$ is analytic in $|t| < c(s')^{-1}$ in the variable t alone with s' as a parameter. Unfortunately, this information on the analyticity domain of $A_s(s',t)$ is useless to extend the range of dispersion relations, because as $s' \to \infty$ the domain in t shrinks to zero. So we have to manipulate F in order to extract something out of this domain. The trick which was used by Mandelstam was to construct a function Φ which has the same discontinuity as F along the s-cut and which is analytic inside a domain as big as possible otherwise. Such a function Φ is easily obtained in this particular case and is given by

$$\Phi(s,t) = \frac{1}{\pi} \int_4^\infty \frac{A_s(s', \frac{st}{s'})}{s'-s} ds' .$$

The discontinuity is clearly

$$\int_4^\infty \delta(s-s') A_s(s', \frac{st}{s'}) ds' = A_s(s',t)$$

and the analyticity domain of the integrand is $|st| < C$. Hence if there is no convergency problem, $\Phi(s,t)$ is analytic in the intersection of the cut s-plane and the domain $|st| < C$, so the function $F(s,t) - \Phi(s,t)$ is analytic in the intersection of the initial domain of F and of $|st| < C$, with the s-cut removed. This technique turns out to be extremely powerful for applications because of its simplicity. The generalization of this theorem by Epstein and Glaser no longer necessitates the explicit construction of Φ :

<u>Theorem of the removal of cuts:</u>

Let $F(z_1, \dots, z_n)$ be analytic in the intersection of a domain D with $\mathbb{C}^{n-1} \times \hat{C}(x_0)$, ($\hat{C}(x_0)$ = cut plane with cut $z \geq x_0$) . Set $z^{(n-1)} = (z_1 \cdots z_n)$. Let $\Delta F(z^{(n-1)}, x_n)$ be the discontinuity of $F(z^{(n-1)}, z_n)$ across this cut. For fixed x_n $F(z^{(n-1)}, x_n)$ is supposed to be analytic in the variable $z^{(n-1)}$ inside a domain δ_{x_n}, the size of this domain may vary with x_n.

Suppose there exists a domain of holomorphy $\Delta \subset \mathbb{C}^n$ such that

$$\Delta \cap \{ z \in \mathbb{C}^n \mid \mathrm{Re}\, z_n = x_n \} \subset \delta_{x_n} .$$

Then one can construct a function $\Phi(z_1 \cdots z_n)$ analytic in the intersection of Δ and $\mathbb{C}^{n-1} \times \hat{C}(x_0)$ which has the same discontinuity as $F(z_1 \cdots z_n)$ across the cut. Hence $F - \Phi$ is analytic in $D \cap \Delta$, which can be used as a starting point for completions. More precisely $F - \Phi$ is analytic in

$\mathcal{H}(D \cap \Delta)$. Of course since $D \cap \Delta \subset \Delta$ and $\mathcal{H}(\Delta) = \Delta$ we have $\mathcal{H}(D \cap \Delta) \subset \Delta$.

Therefore Ψ is analytic in

$$\mathcal{H}(D \cap \Delta) \cap \left(\mathbb{C}^{n-1} \times \hat{\mathbb{C}}(x_0)\right).$$

<u>Remarks:</u> (1) All one needs to know is Δ . It is not necessary to construct ϕ explicitly.

(2) The theorem applies also to the case where several cuts are present.

(3) Δ is not uniquely determined by δ_{x_n} . It is reasonable to choose Δ as big as possible.

(4) In the optimum case

$$\mathcal{H}(D \cap \Delta) = \Delta$$

which is what happened in the case treated by Mandelstam.

(5) In the example we gave

$$\Delta \quad \text{is} \quad |st| < c$$

$$\delta_{s'} \quad \text{is} \quad |t| < \frac{c}{s'} \; .$$

Let us apply this to the π-π - scattering amplitude. The domain $\Delta = \{(s,t) \mid |st| < 256\}$ is a natural domain and for fixed $s > 4$ the section $|t| < 256 s^{-1}$ is contained in the analyticity domain of $A_s(s,t)$. This is because $t = 256 s^{-1}$ is just the right extremity $\gamma_L(s)$ of the large Lehmann ellipse. Similarly the section $|s| < 256 t^{-1}$ is contained in the analyticity domain of $A_t(t,s)$.

Finally the fixed u - sections are also contained in the analyticity domain of $A_u(u,t)$: Indeed, the section is $|u+t-4|\,|t| < 256$. For $\mathrm{Re}\, t > 4$ this is contained in $|u|\,|t| < 256$. Similarly, for $\mathrm{Re}\, s > 4$ it is contained in $|u|\,|s| < 256$. Now on the ellipses of analyticity of A_u with foci $s = 0$ and $t = 0$, $|s| + |t|$ is a constant, and $|s|\,|t|$ is always larger than its value at the extremities of the major axis. Since the extremities of our domain are inside that of the ellipse of analyticity of A_u , the domain is entirely contained in the analyticity domain of A_u . So by the above theorem we can remove all the cuts of $\Psi(s,t,u)$ by subtracting a function ϕ , analytic in $|st| < 256$, which has the same cuts with the same discontinuities. The explicit form of Φ is not important, as mentioned. Moreover, $\mathcal{F}(s,t,u)$ is analytic in $\{|t| < 4\} \otimes \{\text{cut s-plane}\}$ and also in $\{|s| < 4\} \otimes \{\text{cut t-plane}\}$ So in particular $\mathcal{F} - \Phi$ is analytic in

$$\{(|s| < 4) \cup (|t| < 4)\} \cap \{|st| < 256\}$$

This domain is a tube if we choose the variables $\sigma = \log s$ and $\tau = \log t$:

$$\{(\mathrm{Re}\,\sigma < \log 4) \cup (\mathrm{Re}\,\tau < \log 4)\} \cap \{\mathrm{Re}\,\sigma + \mathrm{Re}\,\tau < \log 256\}$$

The tube theorem then gives analyticity in

$$\mathrm{Re}\log s + \mathrm{Re}\log t < \log 256$$

i.e. $\quad |st| < 256$

so $\mathcal{F}$ is analytic in $|st| < 256$ minus the cuts.

Next we want to make use of the results of X (9), from which it is not difficult to see that the domain of holomorphy

$$\Delta_1^{\varepsilon} = \{(s,t) \mid |s-4+\varepsilon||t-16| < 64,\ |s-4+\varepsilon| < 4+\varepsilon\}$$

for $\varepsilon > 0$ arbitrary small, has real sections for s, t, u on the physical cuts. These sections are contained in the analyticity domains of the absorptive part: For the s-channel the first equality, for the t-channel the last equality of X (9) has to be used. For the u -channel we have the conditions

$$u \geq 4$$

$$|t+u+\varepsilon|\,|t-16| < 64$$

$$|t+u+\varepsilon| < 4+\varepsilon \quad (\text{resp. } |s-4+\varepsilon| < 4+\varepsilon)$$

This gives $\mathrm{Re}\,s \geq -2\varepsilon$, $\mathrm{Re}\,t \leq 0$, $|\mathrm{Im}\,t| \leq 4$. This is seen to lie inside the ellipse with foci $t = 0$ and $t = 4-u$ (i.e. $s = 0$) and right extremity $t = 4$.

On the other hand, we have just obtained the result that $\mathcal{F}(s,t,u)$ is analytic in $|st| < 256$ minus the cuts. This domain contains

$$\Delta_2^{\varepsilon} = \{(s,t) \mid |s-4+\varepsilon| < 4,\ |t-16| < 16\}$$
$$\cup \{(s,t) \mid |s-4+\varepsilon| < \varepsilon\}$$

minus the cuts.

This is easily seen by noting that $|s-4+\varepsilon| < 4$ gives $|s| \leq 8$ and $|t-16| < 16$ implies $|t| < 32$. Therefore $\mathbb{F}$ is analytic in the holomorphic envelope of $\Delta_1^{\varepsilon} \cap \Delta_2^{\varepsilon}$ minus the cuts. But then by the tube theorem, $\mathbb{F}$ is analytic in

$$\Delta = \{(s,t) \mid |s-4||t-16| < 64,\ |s-4| < 4\}$$

minus the cuts.

A similar result holds by interchanging s and t. Thus we arrive at the result that for $|s-4| < 4$ and $|t-4| < 4$ the border we got contains the border of the Mandelstam spectral function. There is a gap from $s = 8,\ t = 32$ to $s = 32,\ t = 8$ where we only can continue up to $|st| < 256$. Altogether the analyticity domain for the full amplitude is given by combining the domain $\mathcal{D}'^{\pi\pi}$ of fixed dispersion relations, the domain given by VIII (5) and Δ.

In particular we see that for very low energies, it is a good approximation to consider the analyticity domain as the whole cut plane in the cos Θ or t variable. There is clearly a continuous transition from $s = 4+\varepsilon$, where we have this very large region of analyticity and $s = 4-\varepsilon$, where we have the full cut plane. If the low energy 2π - scattering admits a potential description, the only acceptable potentials will be those which give analyticity in the cut t-plane. The only such potentials are the superpositions of the Yukawa potential plus those potentials decreasing faster than any exponential. However, the description should be just as good for s slightly below $4\mu^2$, where we have at most 2 subtractions and this leaves the Yukawa superpositions being the only admissible potentials.

Next we shortly touch the question of the analyticity domain of partial waves, or more precisely, that of fixed angle analyticity in s of $\mathbb{F}(s, \cos\Theta_s)$ $(-1 \leq \cos\Theta_s \leq 1)$ Now the points $\cos\Theta_s = \pm 1$ give analyticity in the whole cut plane and for cos Θ_s very close to ± 1 the analyticity domain will still be large. Practical calculations show that it is sufficient to discuss the 90° scattering amplitude:

$$\cos\Theta_s = 0\ ;\ \text{i.e}\quad u = t = \tfrac{1}{2}(s-4).$$

So we shall concentrate on domains symmetrical in u and t and once the completion is made, we will take the section $u = t$. In the following, we will employ three kinds of techniques. One is the Mandelstam method with domains of form $|u-\lambda||t-\lambda| \leq c_\lambda$ As shown above this method has the advantage of removing the cuts. For the amplitude itself it even gives new real points of analyticity. The second method will use the fact that when t is inside $\mathcal{D}'^{\pi\pi} \cap \{t \mid \operatorname{Im} t > 0\}$, $F(s,t,u)$ is analytic in $\operatorname{Im} \breve{u}$ and conversely (with u and t interchanged). It will provide points of analyticity in the complex plane but no new real points. The last technique uses results of Bros, Epstein and Glaser.

α) The Mandelstam Method: We still restrict ourselves to the case where λ is in the open interval $(-28, 4) \subset \mathcal{D}'^{\pi\pi}$. Then for sufficiently small ε $F(s,t,u)$ is certainly analytic in $|t-\lambda| < \varepsilon$ and $|u-\lambda| < \varepsilon$ minus cuts. If we succeed in finding a c_λ such that the sections of the domain

$$\Delta_{\lambda, c_\lambda} = \{(s,t,u) \mid |u-\lambda||t-\lambda| < c_\lambda\}$$

with the physical s,t and u cuts are contained in the analyticity domain of A_s, A_t and A_u respectively, then the tube theorem gives analyticity in $\Delta_{\lambda, c_\lambda}$, minus the cuts.

Such determinations for c_λ have been made on a CERN computer. The result is that for λ equal to 0, - 4, - 8, - 12, - 16, - 20, - 24 the corresponding optimal c_λ are 256, 362, 319, 292, 281, 281.7, 158. If we now restrict ourselves to $u = t$, we obtain the condition

XI (1) $$|s - 4 + 2\lambda| < 2 c_\lambda^{\frac{1}{2}}$$

For a given s inside this region consider the straight line in the t-plane connecting $t = \lambda$ and $t = 4 - s - \lambda$ (i.e. $u = \lambda$). Any point of this complex path may be parametrized in the following way:

XI (2) $$\begin{aligned} t &= \alpha\lambda + (1-\alpha)(4-s-\lambda) ; \\ u &= \alpha(4-s-\lambda) + (1-\alpha)\lambda ; \end{aligned} \qquad 0 \leq \alpha \leq 1 .$$

So we have along this path

$$|t-\lambda||u-\lambda| = \alpha(1-\alpha)|4-s-2\lambda|^2 \leq \left|\frac{4-s}{2} - \lambda\right|^2 .$$

Then if s is inside XI (1), the whole segment defined by XI (2) is inside the analyticity domain $|t-\lambda||u-\lambda|<C_\lambda$. On the other hand, if s is outside the cuts, the segments $t=0$ to $t=\lambda+i\varepsilon$ and $u=0$ to $u=\lambda+i\varepsilon$ are inside the analyticity domain, since they are inside $\mathcal{D}'^{\pi\pi}$. The result is that whenever XI (1) is satisfied, we can connect the points $t=0$ and $u=0$ by a complex path. Now the partial wave amplitudes are defined by

$$\frac{1}{s-4}\int_{u=0}^{t=0} P_\ell\left(1+\frac{2t}{s-4}\right) F(s,t,u)\,dt .$$

Starting from s real, we can deform the integration path as s becomes complex and use the complex path we have just obtained.

Therefore the partial wave amplitudes are analytic in XI (1) minus the cuts, and more generally in the union of all domains XI (1) if we let λ vary in (-28,4). This is a rather big domain which extends on the real axis from -28 to 78.

β). We want to use the information that whenever t is in $\mathcal{D}'^{\pi\pi}$, the function $F(s,t,u)$ is analytic in a cut plane in u and conversely. The difficulty is that one of the cuts is moving. However, if we restrict ourselves to t in $\mathcal{D}'^{\pi\pi}(t)\cap\{t\,|\,\mathrm{Im}\,t>0\}$ we can safely say that $F(s,t,u)$ is analytic when t is inside this region and $\mathrm{Im}\,u>0$. These statements remain true when t and u are interchanged.

For simplicity, we insert in $\mathcal{D}'^{\pi\pi}(t)\cap\{t\,|\,\mathrm{Im}\,t>0\}$ an arc of a circle

XI (3) $$0<\arg\frac{B-t}{t-A}<\alpha \qquad (-28\leq A<B\leq 4)$$

and similarly in $\mathcal{D}'^{\pi\pi}(u)\cap\{u\,|\,\mathrm{Im}\,u>0\}$

XI (4) $$0<\arg\frac{B-u}{u-A}<\alpha .$$

Since the condition $\mathrm{Im}\,t>0$ may be written as $0<\arg\frac{B-t}{t-A}<\pi$, and similarly for u the conditions for the two domains are

1) $$0<\arg\frac{B-t}{t-A}<\alpha\ ;\ 0<\arg\frac{B-u}{u-A}<\pi$$

2) $$0<\arg\frac{B-t}{t-A}<\pi\ ;\ 0<\arg\frac{B-u}{u-A}<\alpha .$$

If we use as variables $t' = -i \log \frac{B-t}{t-A}$ and $u' = -i \log \frac{B-u}{u-A}$ we see that the union of these domain is a tube. We then construct the convex hull of the base and get as the envelope of holomorphy the domain satisfying the simultaneous conditions (see page 68)

XI (5) $$\arg \frac{B-t}{t-A} + \arg \frac{B-u}{u-A} < \alpha + \pi$$

$$0 < \arg \frac{B-t}{t-A} \quad ; \quad 0 < \arg \frac{B-u}{u-B} .$$

Here again the simplest thing to do is to find an analyticity domain for the scattering amplitude for $\cos \Theta_s = 0$, i.e. $t = u = \frac{1}{2}(s-4)$. This gives

XI (6) $$0 < \arg \frac{B-t}{t-A} < \frac{1}{2}(\alpha + \pi)$$

or

$$0 < \arg \frac{2B-4+s}{4-s+2A} < \frac{1}{2}(\alpha + \pi)$$

which is the arc of a circle which lies in $\operatorname{Im} s < 0$ (In $\operatorname{Im} s > 0$ of course a symmetric region may be found.) The gain over merely taking the union of XI (3) and XI (4) is that α has been replaced by $\frac{1}{2}(\alpha + \pi)$; i.e., we obtain a much bigger region of the s-plane. The question now is again whether we can for any given s in XI(6) connect $t = 0$ and $u = 0$ by a complex path to define partial wave amplitudes, and whether the path is a straight line. It can easily be checked by geometrical methods that for $A \leq 0 < B$ both conditions are fulfilled so that for s in the domain XI (6), $\Psi(s, \cos \Theta_s)$ is analytic in the neighborhood of $-1 \leq \cos \Theta_s \leq 1$. Hence the partial wave amplitude $f_\ell(s)$ is analytic in the domain XI (6).

The new domain we have obtained extends to $|\operatorname{Im} s| = 80\mu^2$ in the complex direction.

The domain which we show below has been obtained using mere refined techniques. The principle of the method remains the same, only the choice of the initial domain changes. The limits of this approach are not known.

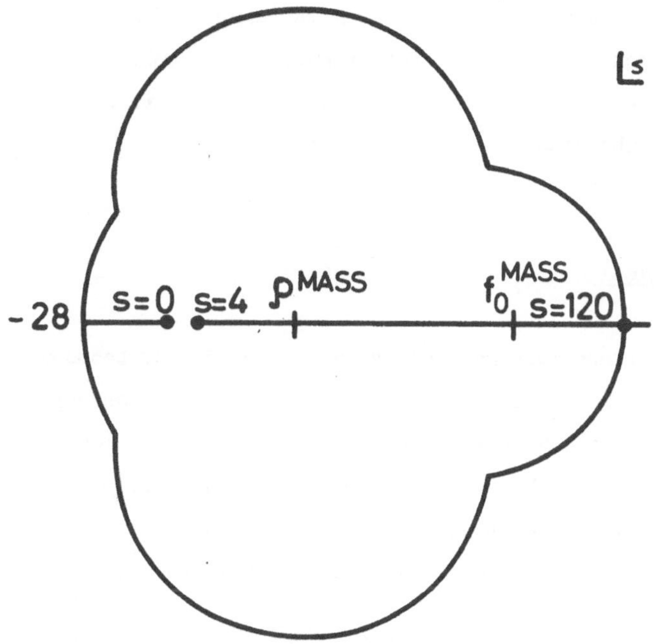

γ)

The final source of information on the analyticity domain of partial wave amplitudes comes from the work of Bros, Epstein and Glaser [B 10]. It is somehow complementary to what we have done up till now, because it provides a domain which extends to infinity in the direction $s \to \infty$. The extension in the complex direction, however, becomes very small.

We do not want to go into details and will only quote the information from which we begin: For fixed negative t_0 the scattering amplitude $\Psi(s, t_0)$ is analytic in s in cones with vertex at s_0 of the form

$$\Phi_{s_0,t_0} > \arg(s-s_0) > 0 \quad ; \quad -\Phi_{s_0,t_0} < \arg(s-s_0) < 0$$

where

$$\cos \Phi_{s_0,t_0} = 1 - 2\left(\frac{32}{36-t_0}\right)^{\frac{\pi}{\Theta_{s_0,t_0}}}$$

and

$$\cos^2 \Theta_{s_0,t_0} = \frac{36-t_0}{s_0} .$$

s_0 can be chosen arbitrary > 0, provided $\frac{36-t_0}{s_0}$.

Finally, we quote another result:

As we already mentioned, Bros, Epstein and Glaser have shown that for any negative t, the scattering amplitude is analytic in $\text{Im } s > 0$ and $\text{Im } s < 0$ minus a finite region of the complex s-plane. We shall say that $\Psi(s,t)$ satisfies a quasi dispersion relation because $\Psi(s,t)$ can be represented as a Cauchy integral over a contour consisting of the physical cuts (from $s = 4-t$ to ∞ and $s = -\infty$ to $s = 0$)

and a contour in the complex plane avoiding the complex singularities from $s = 0$ to $s = 4 - t$. It can be shown that this result is not only valid for t real < 0 but also for any t inside the parabola with focus $t = 0$ and extremity $t = \mu^2$, which of course contains the line t real < 0.

XII. Counterexamples

The preceeding sections may have shown how hard work becomes if one wants to extend the analyticity domain of the scattering amplitude and its absorptive parts. Since we were unable to find the envelope of holomorphy, we would at least like to see how far we may come. Let us consider the $\pi - \pi$ case. We are going to present counterexamples showing that with our present knowledge it is not possible to prove the Mandelstam representation. More precisely we will give an example with the following properties

α) validity of dispersion relations in a domain larger or equal to the one we proved in the preceeding chapters.

β) analyticity in t for $s > 16$ in a domain containing the ellipses previously obtained for the absorptive parts.

γ) a result on crossing of the type Bros, Epstein and Glaser obtained for negative t.

Remark: The counterexample will show validity of dispersion relations also for $4 < s < 16$, therefore the restriction $s > 16$ in β).

The reason that we add dispersion relations in the elastic strip is that otherwise we would have to show the relation between the analyticity domain of the full amplitude and its absorptive part, which is given by elastic unitarity and which was discussed in Chapter IX. This will, however, not be easy to reproduce in a counterexample. A first guide to a counterexample would be

$$\Psi_\nu(s,t,u) = \Psi_\nu(s,t) + \Psi_\nu(s,u) + \Psi_\nu(u,t)$$

with

$$\text{XIII (1)} \quad \Psi_\nu(s,t) = \int_0^1 \int_{16}^\infty \int_{16}^\infty \frac{\rho(M,N,x)\, dM\, dN\, dx}{x(M-s)^\nu + (1-x)(N-t)^\nu} .$$

The quantity $\rho(M,N,x)$ is a measure, subjected only to the condition $\rho(M,N,x) = \rho(N,M,1-x)$ a.e. in order to ensure

$$\Psi_\nu(s,t) = \Psi_\nu(t,s) .$$

For $\nu = \frac{1}{2}$ this is just an unusual way of writing the Mandelstam representation. For if we choose the cuts to be $s \geq 16$ and $t \geq 16$ we see that the conditions $\mathrm{Re}\,(M-s)^{\frac{1}{2}} > 0$ and $\mathrm{Re}\,(N-t)^{\frac{1}{2}} > 0$ give us the cut (s,t) - plane. If the integration were to start at $M = N = 4$, $\nu = 1$ would be the Nakanishi representation which is valid to all orders in perturbation theory [N 1].

The problem is to adjust ν between these two values. We take $\nu = \frac{2}{3}$. For simplicity we drop the integration over M and N and choose M = N = 16 , but all statements are made for the case where we integrate over M and N :

XII (2) $$\Psi(s,t) = \int_0^1 \frac{\rho(x)\, dx}{x(16-s)^{\frac{2}{3}} + (1-x)(16-t)^{\frac{2}{3}}}$$

$0 \neq \rho(x) = \rho(1-x)$ a.e., again such that $\Psi(s,t) = \Psi(t,s)$.

The domain of analyticity is given by

XII (3) $$-\frac{2}{3}\pi \leq \arg(16-s) - \arg(16-t) \leq \frac{3}{2}\pi$$

so the domain is a convex tube in the variables

$$s' = -i \log(16-s) \; ; \; t' = -i \log(16-t).$$

In particular, for $\mathrm{Re}\, t < 16$ we have $|\arg(16-t)| < \frac{\pi}{2}$ and hence this certainly allows $|\arg(16-s)| \leq \pi$ i.e. the cut s-plane. A similar statement holds for s and t interchanged. Graphically we have the following situation for the analyticity domain

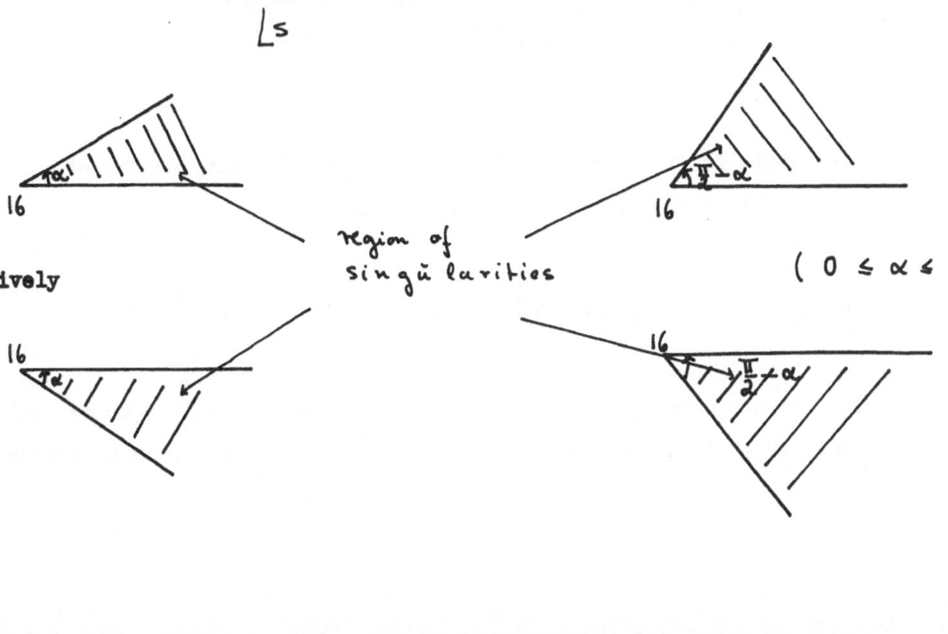

respectively

$(0 \leq \alpha \leq \frac{\pi}{2})$

Now the condition

XII (4) $$-\frac{3}{2}\pi \leq \arg(16-s) + \arg(16-t) \leq \frac{3}{2}\pi$$

gives rise to similar statements. In particular a cut s-plane would be obtained for $\mathrm{Re}\, t < 16$. Graphically

respectively

$(0 \leq \alpha \leq \frac{\pi}{2})$

This is the reason why we also may introduce a term of the form

XII (5) $$G(s,t) = \int_0^\infty \frac{\rho(y)\,dy}{y + (16-s)^{\frac{2}{3}}(16-t)^{\frac{2}{3}}}$$

whose analyticity domain is just given by XIII (4). The analyticity domain of $\Phi(s,t) + G(s,t)$ is then given by the conditions

$$-\frac{3}{2}\pi \leq \arg(16-s) \pm \arg(16-t) \leq \frac{3}{2}\pi$$

In the variables $s' = -i \log(16-s)$ and $t' = -i \log(16-t)$ it is a convex tube, since it is the intersection of two convex tubes. The base is even compact, as is easily seen. Graphically

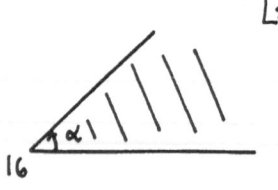

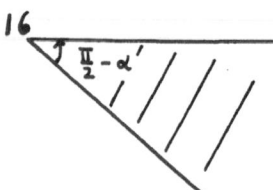

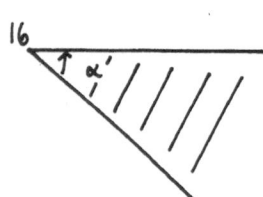

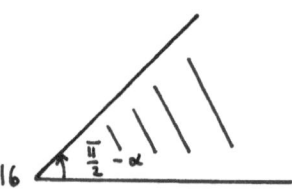

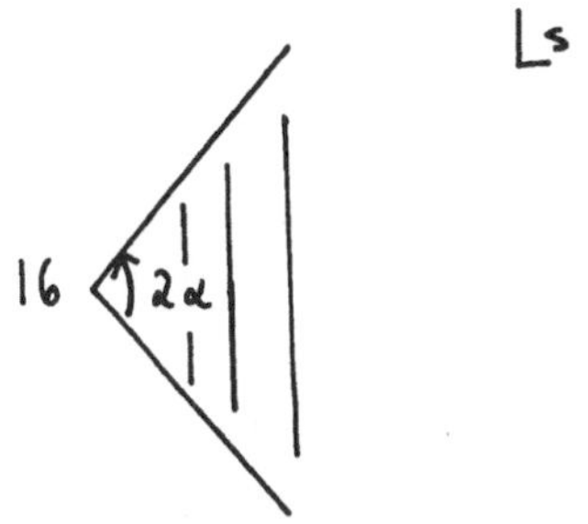

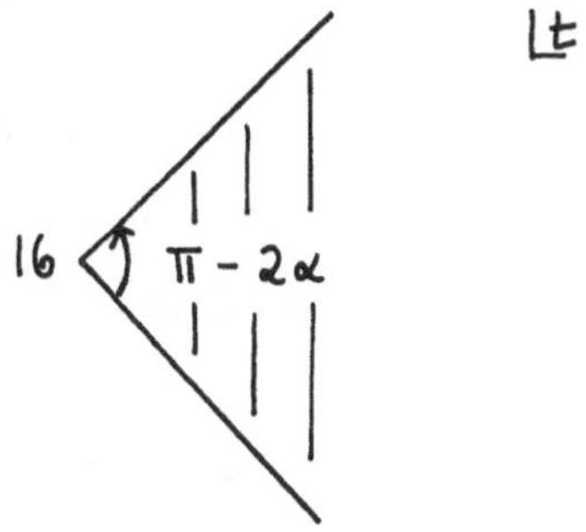

In particular we have analyticity in the cut s-plane (Re t $<$ 16) and vice versa. So we take as our counterexample

XII (6)

$$F(s,t,u) = F(s,t) + G(s,t) + F(u,t) + G(u,t) + F(s,u) + G(s,u).$$

We have to discuss $\alpha)$ $\beta)$ and $\gamma)$:

$\alpha)$: If we do not worry about subtractions, we obviously have fixed t dispersion relations for $\mathrm{Re}\, t < 16$ for

$$F(s,t) + G(s,t) + F(u,t) + G(u,t)$$

since all these terms are analytic in cut planes. The only term to worry about is $F(s,u) + G(s,u)$. However, we have

$$\frac{1}{2}\,(\mathrm{Re}\, u + \mathrm{Re}\, s) = \frac{4 - \mathrm{Re}\, t}{2}$$

so if $\frac{1}{2}(4 - \mathrm{Re}\, t) < 16$ then either $\mathrm{Re}\, s$ or $\mathrm{Re}\, t$ will be less than 16. Thus we will not have any t singularity. Therefore fixed t dispersion relations are valid for $-28 < \mathrm{Re}\, t < 16$.

$\beta)$: We want to discuss the analyticity properties of $F(s,t,u)$ in t for $s = x \pm i0$; $x > 16$.

The term

$$F(s,t) + G(s,t) + F(s,u) + G(s,u)$$

obviously gives the restrictions $\mathrm{Re}\, t < 16$ and $\mathrm{Re}\, u < 16$.

For

$$\mathcal{F}(u,t) + G(u,t)$$

we have no restrictions, since

$$\mathrm{Re}\,u + \mathrm{Re}\,t = 4 - \mathrm{Re}\,s = 4 - x < -12 ,$$

so either Re t or Re u is less than 16 .

Inspection of X (9) shows that we get analyticity in a domain which contains the analyticity domain of the absorptive part of the 2π - scattering amplitude.

γ) : Consider the case $t < -28$

Then

$$\mathcal{F}(s,t) + G(s,t) + \mathcal{F}(u,t) + G(u,t)$$

has only real cuts.

The condition for $\mathcal{F}(s,u)$ to have a singularity is

$$|\arg(16-s)(16-u)^{-1}| \geqslant \tfrac{3}{2}\pi$$

and for $G(s,u)$

$$|\arg(16-s)(16-u)| \geqslant \tfrac{3}{2}\pi .$$

However, inspection of the triangle

$\lfloor s$

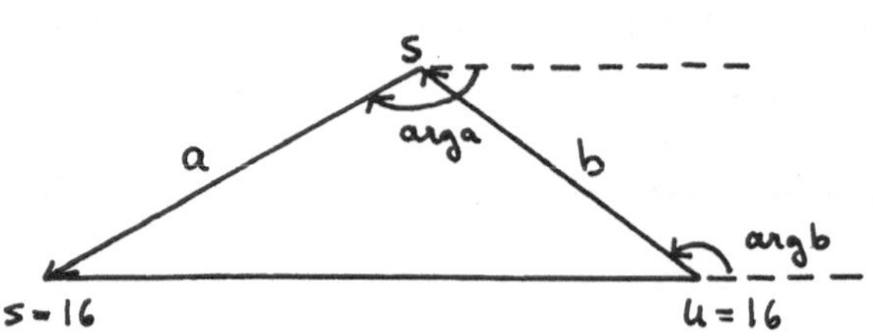

where the vectors a and b are given as

$$a = 16 - s , \quad b = 16 - u$$

shows that only $\mathcal{F}(s,u)$ has singularities. They lie in a circle with a diameter going through s = 16 and u = 16 :

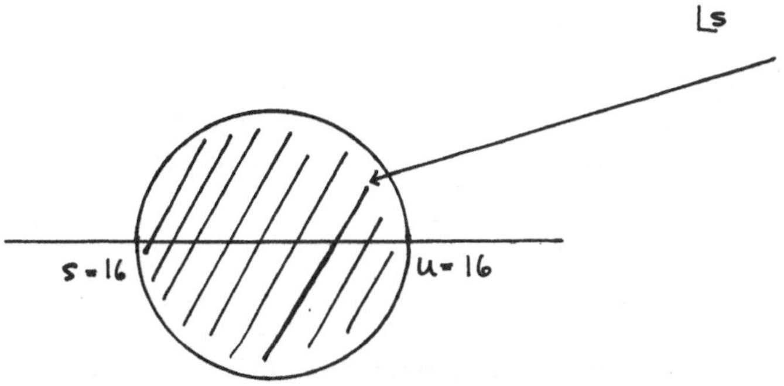

Thus we obtain crossing property just in the form Bros, Epstein and Glaser obtained it with the exception that here we may prove that the extension of the region containing the singularities goes as $|t|$ for $t \rightarrow -\infty$.

Now this example permits us to obtain further results: Consider the more general case $\mathrm{Re}\, t < -28$:

The singularities of $\Psi(s,u)$ are given by two semiarcs of the form

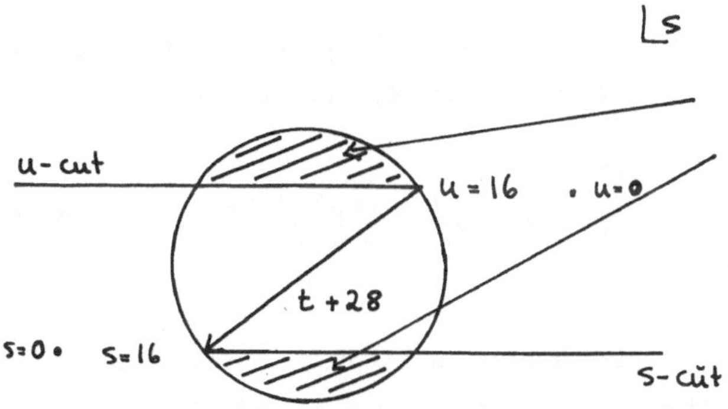

In particular, if $\mathrm{Im}\, t \neq 0$, there exists a path joining $s = 16$ and $u = 16$ and therefore also a path joining $s = 0$ and $u = 0$, not meeting any singularities of

$$\Psi(s,t) + G(s,t) + \Psi(u,t) + G(u,t) + \Psi(s,u).$$

For $G(s,u)$ this is no longer true in general and this is the motivation for introducing the G terms: We distinguish two cases

1) $\frac{\pi}{2} < |\arg(t+28)| < \frac{3}{4}\pi$

2) $\frac{3}{4}\pi < |\arg(t+28)| < \pi .$

In the first case the singularities are of the form

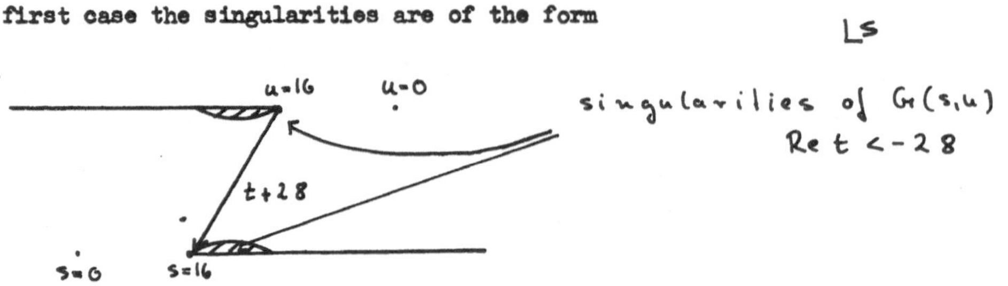

We do not want to describe explicitly the region of singularities. It is sufficient to note that it does not contain the segment joining $s = 16$ and $u = 16$ along which

$$\arg(16-s) = \arg(16-u) = \arg(t+28)$$

such that

$$|\arg(16-u)| + |\arg(16-s)| = 2|\arg(t+28)| < \frac{3}{2}\pi .$$

Therefore we may find a path joining $s = 16$ and $u = 16$ without meeting any singularities of $F(s,t,u)$ and so the partial wave amplitudes in the t-channel

$$f_\ell(t) = \left(\int_{s=0}^{s=16} + \int_{s=16}^{u=16} + \int_{u=16}^{u=0} \right) P_\ell(\cos\theta_t)\, F(s,t,u)\, d\cos\theta_t$$

are analytic in that region. For Re t $>$ 16 we have as the only restrictions $\mathrm{Re}\, s < 16$ and $\mathrm{Re}\, u < 16$: the segment joining $s = 0$ and $u = 0$ is inside the analyticity domain of $F(s,t,u)$. So finally $f_\ell(s)$ is analytic at least in the following domain:

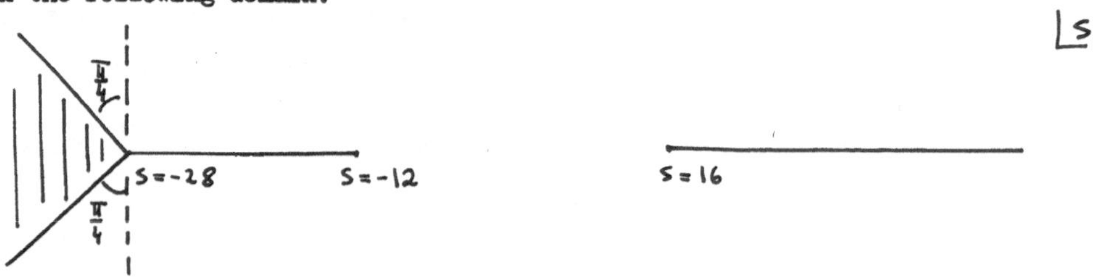

In the second case

$$\frac{3}{4}\pi < |\arg(t+28)| < \pi$$

we have two disconnected regions in the s-plane due to the singularities of $G(s,u)$

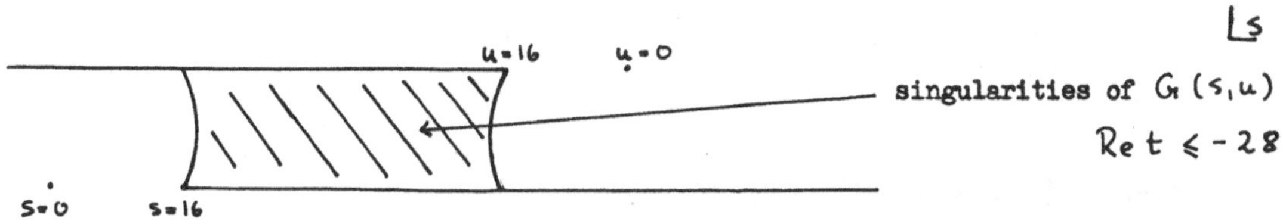

In particular the segment joining $s = 16$ and $u = 16$ is not contained in the analyticity domain of $G(s,u)$. Therefore we cannot improve the above analyticity domain for $f_\ell(s)$.

If we drop the term $G(s,t) + G(s,u) + G(u,t)$ we do not get these singularities which divide the cut s-plane into two disconnected regions. This is normal because $\mathcal{F}(s,t) + \mathcal{F}(s,u) + \mathcal{F}(u,t)$ has a bigger analyticity domain than the perturbation domain of Nakanishi, for which the partial wave amplitudes are analytic in a cut-plane.

More refined examples

We would like to build examples which adhere more closely to the analyticity domain we know, with the hope that they might be used as guidance to get the exact domain of holomorphy of the amplitude.

To do this it is most convenient to interpret geometrically the analyticity domain of our example. The domain of $\mathcal{F}(s,t) + G(s,t)$ can be considered as the envelope of holomorphy of

$$\{(s,t) \mid \mathrm{Re}\, t < 16,\ s \notin [16,\infty)\} \cup \{(s,t) \mid \mathrm{Re}\, s < 16,\ t \notin [16,\infty)\}$$

1st Approximation

To get a smaller domain of validity of dispersion relations, we define $D_{s,t}$ to be the envelope of holomorphy of

$$\Delta_{s,t} = \{(s,t) \mid t \in P_t,\ s \notin [16,\infty)\} \cup \{(s,t) \mid s \in P_s,\ t \notin [16,\infty)\}$$

Here P_t is defined to be the parabola in the complex plane with focus $t = 0$ and extremity $t = 16$, so the condition $t \in P_t$ is equivalent to

$$(\mathrm{Im}\, t)^2 \leq -4 \cdot 16\,(\mathrm{Re}\, t - 16)$$

Choosing the variable $\tau = \tau_1 + i\tau_2 = t^{\frac{1}{2}}$ this is again equivalent to

$$\tau_1 \leq 4$$

i.e. $\pm \arg(\tau - 4) > \frac{\pi}{2}$

Set $\sigma = s^{\frac{1}{2}}$. Then it is easily seen, using the tube theorem, that $D_{s,t}$ is given by the union of 4 disconnected convex tubes of the form

$$\pm \arg(\sigma - 4) \pm \arg(\tau - 4) > \frac{\pi}{2}$$

We note that the rays $\arg(\sigma - 4) = \alpha = \text{const}$ in the σ variable define in the s variable parts of oblique parabolas which have focus O and pass through s = 16. There are no essential qualitative changes but the domain of validity of dispersion relations is smaller. The domain is just

$$D_{s,t} \cap D_{s,u} \cap D_{u,t}$$

For fixed t inside the parabola with focus $t = o$ and extremity $t = 16$, $D_{s,t}$ and $D_{u,t}$ obviously give no singularities except the usual cuts.

We only have to inspect $D_{u,s}$, which is a little bit delicate. There we shall use the following trick: Take first t in the interval (-28, 16). Then

$$\frac{1}{2}(u+s) = c \in (0,16)$$

$$|\arg(16 - s)| + |\arg(16 - u)| <$$
$$|\arg(c - s)| + |\arg(c - u)| = \pi$$

so the point (s,u) is in

$$D'_{s,u} = \{(s,u) \,|\, |\arg(16 - s)| + |\arg(16 - u)| \leq \pi\}$$

which evidently is a connected, convex tube and therefore a domain of holomorphy. We will prove that $D'_{s,u}$ is in $D_{s,u}$. The following argument is used in order to cover also a situation which will appear below. Consider the following "flat" domain:

$$\delta'_{s,u} = \{(s,u) \,|\, s \in (-\infty, 16),\ u \notin [16,\infty)\}$$
$$\cup \{(s,u) \,|\, s \notin [16,\infty),\ u \in (-\infty, 16)\}$$

We have $\delta'_{s,u} \subset \Delta_{s,u}$. Now it may be proved that the envelope of holomorphy of $\delta'_{s,u}$ is just $D'_{s,u}$ which proves that $D'_{s,u} \subset D_{s,u}$ [E 1].

Hence it follows that (modulo subtractions) the following dispersion relations hold for $-28 < t < 16$:

$$\Psi(s,t,u) = \frac{1}{\pi}\int_{16}^{\infty}\frac{A_s(s',t)}{s'-s}\,ds' + \frac{1}{\pi}\int_{16}^{\infty}\frac{A_s(s',t)}{s'-u}\,ds'$$

This expression can be used to make an analytic continuation whenever $A_s(s',t)$ is analytic in t for all $s' > 16$. Therefore it is sufficient to study the analyticity domain of $A_s(s',t)$ and hence of $\Psi(s \pm i0, t)$ for s > 16 . Since the cuts are starting at t = 16 , u = 16 the only limitations are given by t inside the parabola P_t and u inside the parabola P_u for s > 16 . In the t-plane clearly the left parabola P_u moves to the left as s increases and hence the strongest condition arises for s = 16. We therefore get the following domain of validity for fixed t dispersion relations:

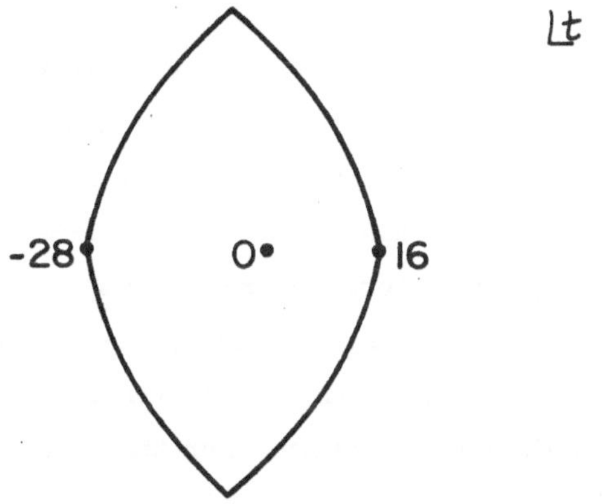

The analyticity in the neighborhood of the physical region is preserved.

Concerning the analyticity domain of partial wave amplitudes we shall content ourselves with looking at the analyticity domain for $\Psi(s, \cos\Theta_s = 0)$. This domain is decisive in deciding whether $\int_{-1}^{+1} \Psi(s, \cos\Theta_s)\, P_\ell(\cos\Theta_s)\, d\cos\Theta_s$ is analytic in s . Then $t = u$ and $D_{u,t}$ gives us a domain limited by the curves

$$\arg(u^{\frac{1}{2}} - 4) = \arg(t^{\frac{1}{2}} - 4) = \pm\frac{\pi}{2}$$

which are just parts of the parabolas

$$(\mathrm{Re}\, u)^2 = \pm 32(\mathrm{Im}\, u + 8)$$

$$(\mathrm{Re}\, t)^2 = \pm 32(\mathrm{Im}\, t + 8)$$

The domains $D_{s,t}$ and $D_{s,u}$ give no restrictions because the relation $s = 4 - 2t = 4 - 2u$ easily shows that the point (s,t) (resp. (s,u)) is contained in $D'_{s,t}$ (resp. $D'_{s,u}$) . Hence the domain of the 90° scattering amplitude and of the partial wave amplitudes is given by

where γ_+ and γ_- are parts of the parabolas

$$\left(\mathrm{Re}\left(\frac{4-s}{2}\right)\right)^2 = \pm 32\left(\mathrm{Im}\,\frac{4-s}{2} + 8\right).$$

2nd approximation

In the previous step we have described a smaller domain of analyticity. However, what we want to demonstrate now is that we can arbitrarily specify the domain of analyticity of the amplitude for s real $>$ 16 and construct a function that for s approaching the interval $[16, \infty)$ has precisely that domain of analyticity and no more. Therefore we shall take as input the analyticity domain of the absorptive part of the 2π - scattering amplitude for $s > 16$ that we have obtained in the previous sections. So if we succeed we will have constructed an amplitude which has exactly the same domain of validity of fixed t dispersion relations as the one we obtained for the 2π - scattering amplitude (see page 65).

We start from a domain $D_{s,t}^{\sigma_1,\sigma_2}$ in which the parabola of $D_{s,t}$ is replaced by $\bigcup_{16<\sigma_1<\sigma<\sigma_2}$, where E_σ is the ellipse of analyticity of the absorptive part (X (9)). What we want is the following: For $s \in [\sigma_1, \sigma_2]$, t should vary in $\bigcup_{\sigma_1\leq\sigma\leq\sigma_2} E_\sigma$ and for $s \notin [\sigma_1, \sigma_2]$, t should be any point the cut plane. The same should be true for s and t interchanged. Hence we define $D_{s,t}^{\sigma_1,\sigma_2}$ to be the analytic completion of

$$\Big\{(s,t) \;\Big|\; t \in \bigcup_{\sigma_1\leq\sigma\leq\sigma_2} E_\sigma \cup (-\infty,\sigma_1) \cup (\sigma_2,\infty),\; s \notin [\sigma_1,\sigma_2]\Big\}$$

$$\cup \Big\{(s,t) \;\Big|\; s \in \bigcup_{\sigma_1\leq\sigma\leq\sigma_2} E_\sigma \cup (-\infty,\sigma_1) \cup (\sigma_2,\infty),\; t \notin [\sigma_1,\sigma_2]\Big\}.$$

Graphically

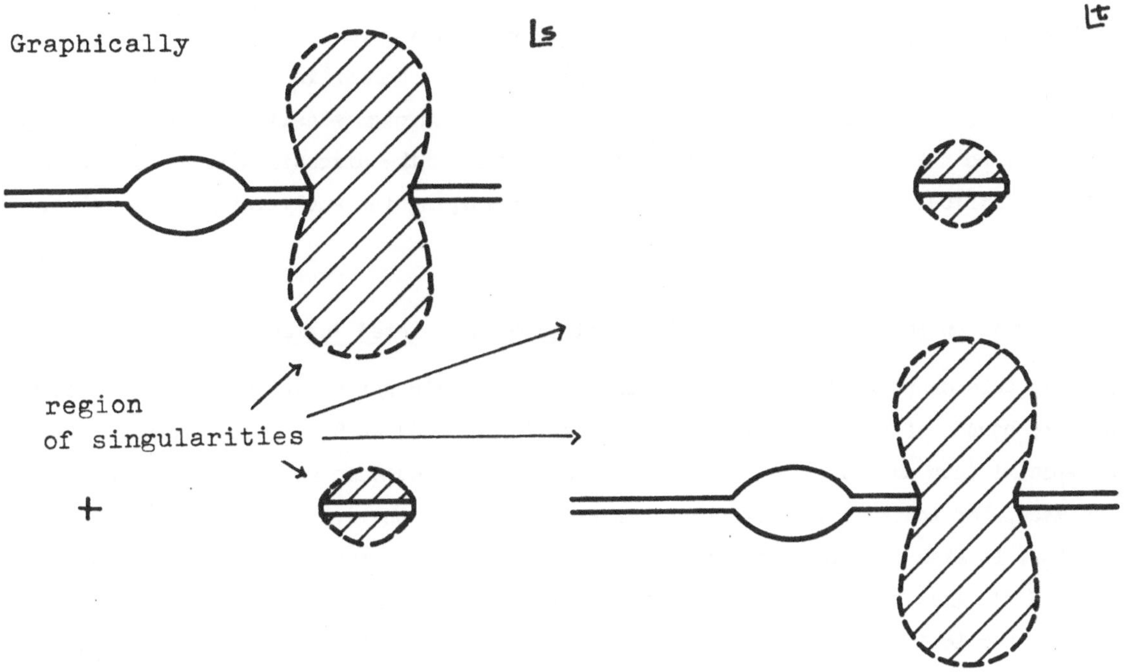

The dotted lines are the interpolations by the completions which can be reduced, after convenient mappings, to making tubes convex.

By construction it is clear that

$$D^{\sigma_1,\sigma_2}_{s,t} \cap D^{\sigma_1,\sigma_2}_{u,s} \cap D^{\sigma_1,\sigma_2}_{u,t}$$

will have sections for real s (resp. t , u) $>$ 16 which are larger or equal to what we have obtained. It is also clear that we can repeat the argument for dispersion relations by slight modifications.

In particular dispersion relations will be valid in a domain larger than the one we obtained for 2π - scattering. Now we can take the intersection of all domains

$$D^{\sigma_1,\sigma_2}_{s,t} \cap D^{\sigma_1,\sigma_2}_{u,s} \cap D^{\sigma_1,\sigma_2}_{u,t} .$$

By taking σ_1 and σ_2 sufficiently close to each other we can have $\bigcup_{\sigma_1 \leq \sigma \leq \sigma_2} E_\sigma$ as close as we wish to E_s with $\sigma_1 \leq s \leq \sigma_2$. Therefore we get for any $s > 16$ the analyticity domain that we had obtained for the absorptive part plus a real segment and nothing more. Hence fixed t dispersion relations will be valid in the domain that has been previously obtained plus the interval (4,16).

Also we may repeat the discussion of analyticity of the partial wave amplitudes, showing that (apart from cuts) the $f_\ell(s)$ are analytic for $\mathrm{Re}\, s > -28$, a result which might be true for the real amplitude. The possibility of reducing the domain of analyticity for $4 < s < 16$ seems very questionable. If the domain is smaller than a cut plane, the domain of analyticity of the absorptive part will through elastic unitarity be bigger and therefore this implies some special restrictions on the discontinuity of F which might be difficult to take into account without enlarging the analyticity of F also.

XIII Bounds on the two pion scattering amplitude and related topics

In this chapter we will discuss some modest results which have been obtained on the absolute bounds of the two pion scattering amplitude [L 4] [L 5] . These bounds are given for special (non physical) values of s, t and u and contain as its only parameter the pion mass. For instance one can obtain an upper bound at the symmetric point $s = t = u = \frac{4}{3}\mu^2$. It also implies that the pion-pion scattering lengths have a negative lower bound [B 7]. Also the physical scattering amplitude is shown to satisfy sum rule inequalities. In particular the total pion-pion cross-section satisfies such a sum rule, which gives sum rules for two pion inelastic cross-sections, for instance for the cross-sections $\pi\pi \leftrightarrow p\bar{p}$ and $\pi\pi \leftrightarrow K\bar{K}$. It is remarkable that no information on the masses of kaons and protons is needed.

These results have of course a dynamical content: They prevent the two pion amplitude having a too wild behaviour. At least physically this is not surprising, since we start from the following facts:

i) The two pion system has no bound states.

ii) The nearest singularities are produced by the exchange of two pions.

i) means that we start with dispersion relations without pole terms. The result ii) was obtained in chapter VI.

Also these results in some way complement the Froissart bound which we obtained as an asymptotic bound, however, without having specified where this asymptotic region should start.

We will treat the case of elastic $\pi^0 - \pi^0 -$ scattering, where complete crossing symmetry holds. Also s, t, u will be restricted to the Mandelstam triangle, where $s, t, u < 4$ and where $F(s,t,\ )$ is a real analytic function.

Of course we will prove all statements without assuming the validity of Mandelstam representation. Even if Mandelstam representation were true, it would not be of any particular help simplifying arguments. Our method will be some kind of an inverse bootstrap technique.

To make the discussion more transparent, let us start with unsubtracted dispersion relations in the Mandelstam triangle $[\mu^2 = 1]$

$$F(s,t,u) = \frac{1}{\pi}\int_4^\infty \frac{A_s(s',t)}{s'-t}\,ds' + \frac{1}{\pi}\int_4^\infty \frac{A_s(s',t)}{s'-u}\,ds' .$$

First we want to estimate $A_s(s,t)$ by means of $F(s,0,4-s)$. More precisely we have

XIII (1) $$A_s(s,t) \geq \phi(s,t)\,|F(s,0,4-s)|^2 \;; \quad 0 < t < 4.$$

$$\phi^{-1}(s,t) = \frac{s^{\frac{1}{2}}}{2k}\sum_{\ell\ \mathrm{even}}\frac{(2\ell+1)}{P_\ell(1+\frac{t}{2k^2})} .$$

For the proof we write

$$|F(4,0,4-s)|^2 = \left|\frac{s^{\frac{1}{2}}}{2k}\sum_{\ell\ \mathrm{even}}(2\ell+1)f_\ell(s)\right|^2$$

$$= \left(\frac{s^{\frac{1}{2}}}{2k}\right)^2\left|\sum_{\ell\ \mathrm{even}}(2\ell+1)^{\frac{1}{2}}f_\ell(s)\,P_\ell\left(1+\frac{t}{2k^2}\right)^{\frac{1}{2}}\cdot\frac{(2\ell+1)^{\frac{1}{2}}}{P_\ell(1+\frac{t}{2k^2})^{\frac{1}{2}}}\right|^2$$

$$\leq \left(\frac{s^{\frac{1}{2}}}{2k}\right)^2\sum_{\ell\ \mathrm{even}}(2\ell+1)\,|f_\ell(s)|^2\,P_\ell\left(1+\frac{t}{2k^2}\right)\cdot\sum_{\ell\ \mathrm{even}}\frac{(2\ell+1)}{P_\ell(1+\frac{t}{2k^2})} ,$$

using Schwarz' inequality. Since unitarity gives $|f_\ell(s)|^2 \leq \mathrm{Im}\, f_\ell(s)$, XIII (1) follows immediately. If $A < \frac{s^{\frac{1}{2}}}{2k}$, XIII (1) is essentially the best estimate obtainable.

Also

XIII (2) $$\sum_{\ell\ \mathrm{even}}^{\infty}\frac{(2\ell+1)}{P_\ell(1+\frac{t}{2k^2})} < \mathrm{const}\cdot\frac{2k^2}{t} ,$$

which may be obtained from

$$P_\ell(x) \geq \frac{\phi_0}{\pi}\left(1+\cos\phi_0\sqrt{x^2-1}\right)^\ell \qquad \text{(see page 37)}$$

$$0 \leq \phi_0 \leq \frac{\pi}{2} \;; \quad 1 < x .$$

and

$$\sum (2\ell+1)\,\alpha^{\ell} \sim \frac{c}{(1-\alpha)^2} \quad ; \quad 0 < \alpha < 1 .$$

If on the other hand $A > \frac{s^{\frac{1}{2}}}{2k}$ we get a better estimate by following the method of page 39 where we found the minimum of $A_s(s,t)$ $(0 < t < 4)$ compatible with unitarity, when

$$\bar{\mathcal{F}}(s) = \frac{s^{\frac{1}{2}}}{2k} \sum_{\ell \text{ even}} (2\ell+1)\,|f_\ell(s)|$$

was given. The result was

XIII (3) a) $f_\ell(s) = 1$ for $\ell = 0, 2, \cdots 2\tilde{L}(s)$

b) $f_\ell(s) = \dfrac{c(s)}{P_\ell\left(1+\frac{t}{2k^2}\right)}$ for $\ell = 2\tilde{L}(s)+2, \cdots$

c) $1 \leq P_{2\tilde{L}(s)}\left(1+\frac{t}{2k^2}\right) < c(s) \leq P_{2\tilde{L}(s)+2}\left(1+\frac{t}{2k^2}\right).$

So $\bar{\mathcal{F}}(s)$ and $\varphi(\bar{\mathcal{F}}, s, t) = A_{min}(s,t)$ may be written in the form

XIII (4)
$$\bar{\mathcal{F}}(s) = \frac{s^{\frac{1}{2}}}{2k}\left\{ \sum_{0}^{2\tilde{L}(s)} (2\ell+1) + c(s) \sum_{2\tilde{L}(s)+2}^{\infty} \frac{(2\ell+1)}{P_\ell\left(1+\frac{t}{2k^2}\right)} \right\}$$

$$\varphi(\bar{\mathcal{F}}, s, t) = \frac{s^{\frac{1}{2}}}{2k}\left\{ \sum_{0}^{2\tilde{L}(s)} (2\ell+1)\,P_\ell\left(1+\frac{t}{2k^2}\right) + c^2(s) \sum_{2\tilde{L}+2}^{\infty} \frac{(2\ell+1)}{P_\ell\left(1+\frac{t}{2k^2}\right)} \right\}$$

with the condition for $c(s)$

$$P_{2\tilde{L}(s)}\left(1+\frac{t}{2k^2}\right) < c(s) \leq P_{2\tilde{L}(s)+2}\left(1+\frac{t}{2k^2}\right)$$

For any fixed s and t, $\varphi(\bar{\mathcal{F}}, s, t)$ may be shown to be increasing when $\bar{\mathcal{F}}$ increases. Thus we have, since $|\mathcal{F}(s,\cos\theta)| \leq \bar{\mathcal{F}}(s)$ for physical $\cos\theta$:

XIII (5) a) $$A_s(s,t) \geq \varphi(\overline{\Psi}, s, t) \geq \varphi(|\Psi(s,0,4-s)|, s, t) ; s \geq 4$$

b) $$A_s(s,t) \geq \varphi(|\Psi(s, 4-\bar{s}-t, \bar{s}+t-s)|, s, t)$$

for $s \geq \bar{s} + t > 4$; $0 < t < 4$.

The inequality b) cannot be obtained for $4 < s < \bar{s} + t$; in this region the amplitude $\Psi(s, 4-\bar{s}-t, \bar{s}+t-s)$ is unphysical.

For fixed s and large $\overline{\Psi}$, XIII (4) leads to the explicit relation [L 5] :

XIII (6) $$A_s(s,t) \underset{\substack{\overline{\Psi} \to \infty \\ s,t \text{ fixed}}}{\cong} \sqrt{\frac{\sqrt{2\overline{\Psi}}}{\pi}} \exp\left\{\sqrt{2\overline{\Psi}} \log \alpha\right\} \cdot$$

$$\cdot \frac{\exp\left[-\log \alpha \cdot \left\{\frac{2\alpha^2}{\alpha^2-1} - \frac{1}{2}\right\}\right]}{\sqrt{\alpha^2-1}\;\sqrt{x^2-1}} \cdot$$

$$\cdot \left(1 + \frac{2\alpha^3}{(\sqrt{x+1} + \sqrt{x-1})^2}\right) ,$$

$$\alpha = x + \sqrt{x^2-1} \quad ; \quad x = 1 + \frac{t}{2k^2} .$$

First we shall use the simpler inequality XIII(1) and insert it into the dispersion relation

XIII (7) $$\Psi(s,t,u) \geq \frac{1}{\pi}\int_4^\infty \Phi(s',t)\left\{\frac{1}{s'-s} + \frac{1}{s'-u}\right\}|\Psi(s',0,4-s')|^2 ds'.$$

Next we want to find a lower bound of the right hand side of XIII (7) by something proportional to $|\Psi(s_0, 0, 4-s_0)|^2$, where s_0 is to be chosen conveniently.

To motivate the following, let us consider a simple example:

Let $g(z)$ be a real, analytic function in the cut z-plane with cut $1 \leq z < \infty$

Let $\omega(x)$ be a positive measurable function on $[1, \infty)$ and set

$$C(g) = \int_1^\infty \omega(x)|g(x)|^2 dx$$

Now it is reasonable to ask: Does there exist $\chi(\omega)$ such that

$$\text{XIII (8)} \qquad |g(0)|^2 \leq \chi(\omega)\, C(g)$$

for a suitable family of analytic functions $g(z)$ of the form just described? Equation XIII (8) would then be something like an averaged maximum modulus theorem of the same type as Poisson's inequality. This problem has been discussed by Meiman [M 13] in connection with attempts to estimate coupling constants in field theory [G 1] .

He showed that it suffices to postulate

$$|g(z)| < C_\varepsilon \exp\left(\varepsilon |z|^{\frac{1}{2}}\right)$$

for all $\varepsilon > 0$ and suitable C_ε .

The trick used was to map the cut plane into the unit circle and then apply results of Smirnow and Szegö [S 3] [S 10] on forms

$$\int_{-\pi}^{+\pi} p(\theta)\, |f(e^{i\theta})|^2 \, d\theta$$

for functions f analytic in the unit disc. $p(\theta)$ is a positive measurable function, which in our example is determined by $\omega(x)$. The condition on p (and therefore also on ω) is

$$\left| \int_{-\pi}^{+\pi} \ln p(\theta)\, d\theta \right| < \infty .$$

We will imitate this procedure for the right hand side of XIII (7). First $\Psi(s, 0, 4-s)$ is analytic in the double cut plane with cuts $s \leq 0$ and $s \geq 4$. By the transformation

$$\text{XIII (9)} \qquad z = \frac{(s-2)^2}{4}$$

we obtain a cut plane with a cut $1 \leq z < \infty$. Next we transform our z-plane to a unit disc $|y| \leq 1$ by the mapping z $\longrightarrow$ y

$$\text{XIII (10)} \qquad y = -\frac{(\zeta - i)}{(\zeta + i)} \; ; \; \zeta = \sqrt{\frac{z-1}{1-c}}$$

$$c = z(s_0) = \frac{(s_0 - 2)^2}{4} .$$

As mentioned, s_0 will be chosen later. Note that the image of $s = s_0$ is $y = 0$. The integration range of s', $s' \geq 4$, is mapped onto the upper half circle and we may write XIII (7) as

$$\text{XIII (11)} \qquad \Psi(s,t,u) \geq \frac{1}{2\pi} \int_{-\pi}^{+\pi} W_{s_0}(\alpha,s,t)\, |\Psi(s(\alpha),0,4-s(\alpha))|^2\, d\alpha .$$

Here $W_{s_0}(\alpha,s,t)$ is a positive measure which contains the Jacobian of the transformation and otherwise is given by

$$\phi(s(\alpha),t) \left\{ \frac{1}{s(\alpha)-s} + \frac{1}{s(\alpha)-u} \right\} \geq 0 .$$

Also we have used the fact that

$$\Psi^*(s^*,0,4-s^*) = \Psi(s,0,4-s) \qquad \text{for general } s$$

implies

$$\Psi^*(s(-\alpha),0,4-s(-\alpha)) = \Psi(s(\alpha),0,4-s(\alpha))$$

by which the integration may be extended to the whole circle.

Next we use a simple trick to get rid of the weight function. The theorem on arithmetical and geometrical means [S 10] gives

$$\frac{1}{2\pi} \int_{-\pi}^{+\pi} W_{s_0}(\alpha,s,t)\, |\Psi(s(\alpha),0,4-s(\alpha))|^2\, d\alpha$$

$$\geq \exp \frac{1}{2\pi} \int_{-\pi}^{+\pi} \log \{ W_{s_0}(\alpha,s,t)\, |\Psi(s(\alpha),0,4-s(\alpha))|^2 \} d\alpha$$

$$= \exp \frac{1}{2\pi} \int_{-\pi}^{+\pi} \log W_{s_0}(\alpha,s,t)\, d\alpha \; \exp \frac{1}{\pi} \int_{-\pi}^{+\pi} \log |\Psi(s(\alpha),0,4-s(\alpha))|\, d\alpha$$

In the first factor, we may transform back to the variable s' again and for the second we apply Poisson's inequality for subharmonic functions ρ :

$$\rho(0) \leq \frac{1}{2\pi} \int_{-\pi}^{\pi} \rho(e^{i\theta})\, d\theta .$$

This leads to the inequality

XIII (12) $$\frac{1}{2\pi}\int_{-\pi}^{+\pi} W_{s_0}(\alpha, s, t)\,|\mathcal{F}(s(\alpha), 0, 4-s(\alpha))|^2\, d\alpha$$

$$\geqslant \exp \int_4^{\infty} J_{s_0}(s') \log J_{s_0}(s')^{-1} \cdot \log\left\{\Phi(s',t)\left(\frac{1}{s'-s}+\frac{1}{s'-u}\right)\right\} ds'$$

$$\cdot\, |\mathcal{F}(s_0, 0, 4-s_0)|^2$$

where $J_{s_0}(s')$ is the Jacobian of the transformation. Combined with XIII (11) we finally obtain

$$\mathcal{F}(s,t,u) \geqslant H(s,t,s_0)\,|\mathcal{F}(s_0, 0, 4-s_0)|^2$$

where $H(s,t,s_0)$ is explicitly given through XIII (12). If we now choose $s = t = s_0 = 2$ and use crossing, we arrive at

XIII (13) $$\mathcal{F}(2,0,2) \geqslant H(2,2,2)\,\mathcal{F}(2,0,2)^2$$

such that

$$0 \leqslant \mathcal{F}(2,0,2) \leqslant H(2,2,2)^{-1}.$$

Numerial calculations give

$$0 \leqslant \mathcal{F}\left(\tfrac{4}{3}, \tfrac{4}{3}, \tfrac{4}{3}\right) \leqslant \mathcal{F}(2,0,2) \leqslant 1.8 \;.$$

In particular, since

$$A_s(s,0) = \frac{k \cdot s^{\frac{1}{2}}}{8\pi}\, \sigma_{tot}^{\pi_0\pi_0}(s)$$

we obtain the following sum rule

$$\frac{2}{\pi}\int_4^{\infty} \frac{k \cdot s^{\frac{1}{2}}}{4\pi}\, \sigma_{tot}^{\pi_0\pi_0}(s)\, \frac{1}{s-2}\, ds \leqslant 3.6$$

Now let us turn to the case of dispersion relations with subtractions. We know that we need at most two subtractions. Due to the crossing symmetry $u \leftrightarrow s$ only one subtraction constant $C(t)$ appears. Thus we may write

XIII (14) $$\mathcal{F}(s_1, t, u_1) - \mathcal{F}(s_2, t, u_2)$$

$$= \frac{(s_1 - s_2)(s_1 - u_2)}{\pi} \int_4^\infty A_s(s', t) \frac{(2s' - 4 + t) \quad ds'}{(s' - s_2)(s' - s_1)(s' - u_2)(s' - u_1)}$$

with $u_1 + s_1 = u_2 + s_2 = 4 - t.$

It is important to notice that if

$$-t < s_2 < 4 \; ; \; -t < s_1 < 4 \quad ; \quad 0 < t < 4$$

the integrand in XIII (14) has a definite sign. If in addition $s_1 > u_2$, $s_1 > s_2$ then $\mathcal{F}(s_1, t, u_1) - \mathcal{F}(s_2, t, u_2) > 0$. We will choose $u_2 = 0$, so $s_2 = 4 - t$. Now we want to repeat our old arguments. But since we start from two points (s_1, t, u_1) and (s_2, t, u_2), we will need two estimates. As the integrand in XIII (14) is positive, we may at once obtain an estimate using previous arguments. However, instead of using XIII (1), we will now concentrate on the estimates XIII (5). Making the same change of variables as in XIII (9) and XIII (10), we get a first inequality similar to XIII (11):

XIII (15) $$\mathcal{F}(s_1, t, u_1) - \mathcal{F}(4 - t, t, 0)$$

$$\geq \frac{1}{2\pi} \int_{-\pi}^{\pi} W_{s_0}(\alpha)\, \varphi(|\mathcal{F}(s(\alpha), 0, 4 - s(\alpha))|, s(\alpha), t)\, d\alpha$$

where $W_{s_0}(\alpha)$ is a known function of α. Also we did not write out explicitly the dependence on s_1. On the other hand we have Poisson's inequality

XIII (16) $$|\mathcal{F}(s_0, 0, 4 - s_0)| \leq \exp \frac{1}{2\pi} \int_{-\pi}^{\pi} \ln |\mathcal{F}(s(\alpha), 0, 4 - s(\alpha))|\, d\alpha.$$

Now equation XIII (15) with the left hand fixed does not allow for unlimited growth of $|\mathcal{F}(s(\alpha), 0, 4 - s(\alpha))|$. As a result we obtain a maximum for the right hand side of XIII (16) with subsidiary condition XIII (15). This maximum gives via XIII (16) an upper bound for $|\mathcal{F}(s_0, 0, 4 - s_0)|$. Using a Lagrange multiplier λ, this condition may be written as

XIII (17) $$W_{s_0}(\alpha)\, |\mathcal{F}(s(\alpha), 0, 4 - s(\alpha))|\, \frac{\partial \varphi}{\partial |\mathcal{F}|} (|\mathcal{F}(s(\alpha), 0, 4 - s(\alpha))|, s(\alpha), t)$$

$$= \lambda = \text{const.}$$

It may be shown that equations XIII (4) imply the increase of

$$\varphi(|\Phi|, s(\alpha), t) \quad \text{and} \quad |\Phi| \frac{\partial \varphi}{\partial |\Phi|}(|\Phi|, s(\alpha), t)$$

with $|\Phi|$ for arbitrary fixed α and t.

Therefore if we increase λ in XIII (17) then also $|\Phi(s(\alpha), 0, 4-s(\alpha))|$ increases for any α. But because $\varphi(|\Phi|, s(\alpha), t)$ increases with $|\Phi|$, it may be shown that λ cannot exceed some critical value λ_{max} and that $|\Phi(s(\alpha), 0, 4-s(\alpha))|$ corresponding to λ_{max} gives through XIII (16) an upper bound on $|\Phi(s_0, 0, 4-s_0)|$. If we set $s_0 = 4-t$ and use crossing we finally get an inequality of the form

XIII (18)
$$\Phi(s_1, t, u_1) - \Phi(4-t, t, 0) \geqslant H_1(\Phi(4-t, t, 0), s_1, t) .$$

To get the second inequality we will use XIII (5) b) for $s_1 = \bar{s}$. We make a change of variables:

XIII (19)
$$z = \frac{(2s - s_1 - t)^2}{(8 - s_1 - t)^2}$$

and

XIII (20)
$$y = -\frac{(\zeta - i)}{(\zeta + i)} \quad ; \quad \zeta = \left(\frac{z - a}{a - c}\right)^{\frac{1}{2}}$$

where
$$a = z(s = s_1 + t) = \frac{(s_1 + t)^2}{(8 - s_1 - t)^2} > 1$$

$$c = z(s = s_1) = \frac{(s_1 - t)^2}{(8 - s_1 - t)^2} .$$

Thus $s = s_1 \to y = 0$ and $s = s_1 + t \to y = 1$.
For $s > s_1 + t$, $|y(s)| = 1$ so that the part of the cut corresponding to the integration region $(s_1 + t, \infty)$ has been transformed onto the upper half circle. In analogy with XIII (18)

XIII (21)
$$\Phi(s_1, t, u_1) - \Phi(4-t, t, 0) \geqslant \frac{1}{2\pi} \int_{-\pi}^{+\pi} \hat{W}_{s_1}(\alpha)\, \varphi(|\Phi_1(\alpha)|, \alpha)\, d\alpha$$

where

$$\Psi_1(\alpha) \underset{\text{def.}}{=} \Psi(s(\alpha),\ 4-s_1-t,\ s_1+t-s(\alpha))$$

$$\varphi(|\Psi_1(\alpha)|,\alpha) \underset{\text{def.}}{=} \varphi(|\Psi_1(\alpha)|,\ s(\alpha),t).$$

The rest of the cut , i.e. $(4, s_1+t)$, is mapped onto the interval $(\beta, 1)$ in the y-plane where

XIII (22) $$0 < \beta = y(s=4) = \frac{(s_1 t)^{\frac{1}{2}} - 2(s_1+t-4)^{\frac{1}{2}}}{(s_1 t)^{\frac{1}{2}} + 2(s_1+t-4)^{\frac{1}{2}}} < 1.$$

Since $(\beta, 1)$ is a cut for *)

$\Psi(s(y),\ 4-s_1-t,\ s_1+t-s(y))$ in y , we must look for a substitute of XIII (16). This is given by the following estimate

XIII (23) $$|\Psi(s_1, 4-s_1+t, t)| \leq \frac{1}{\beta} \exp \frac{1}{2\pi} \int_{-\pi}^{+\pi} \ln |\Psi(s(\alpha), 4-s_1-t, s_1+t-s(\alpha))| \, d\alpha.$$

Equation XIII (23) is a consequence of the following lemma:

<u>Lemma:</u> Let $f(y)$ be a real analytic function in the unit disc with a cut $\beta \leq y \leq 1$ $(0<\beta<1)$ such that $\mathrm{Im}\, f > 0$ on the upper edge of the cut. Then

$$|f(0)| \leq \frac{1}{\beta} \exp \frac{1}{2\pi} \int_{-\pi}^{+\pi} \ln |f(e^{i\theta})| \, d\theta.$$

It is easily seen that the conditions of the lemma are satisfied in our case. Remember that $s_1 + t - s(y) > 0$ if $\beta \leq y \leq 1$, so we may use crossing and positivity (see the statements connected with V (14)).

<u>Proof of the Lemma:</u> We may obviously assume $f(0) \neq 0$. Let

$$z_j = r_j e^{i\varphi_j} \quad (0 < r_j < 1) \quad (1 \leq j \leq n)$$

be the (not necessarily different) zeros of $f(y)$ and set

---

*) Since we write $s = s(y)$, to be consistent, we should also have written $s(e^{i\alpha})$ instead of $s(\alpha)$. We hope this gives no confusion.

$$f(y) = \prod_{j=1}^{n} \frac{z - r_j e^{i\varphi_j}}{- r_j e^{i\varphi_j}} \tilde{f}(y)$$

Then $\tilde{f}(0) = f(0)$, $\tilde{f}(y)$ has no zeros and it is easy to see that $\operatorname{Im} \tilde{f} > 0$ on the upper edge of the cut. Applying Cauchy's formula for we get

$$\ln \tilde{f}(0) = \frac{1}{2\pi} \int_{-\pi}^{\pi} \ln \tilde{f}(e^{i\theta})\, d\theta$$

$$+ \frac{1}{2\pi i} \int_{\beta}^{1} \frac{\ln \tilde{f}(x+i0) - \ln \tilde{f}(x-i0)}{x}\, dx$$

If we set $\tilde{f}(y) = |\tilde{f}(y)| e^{ie(y)}$ then, because of the assumption $\operatorname{Im} f > 0$ we have

$$0 \leq e(x+i0) = -e(x-i0) \leq \pi \quad (\beta \leq x \leq 1).$$

Hence

$$\ln|\tilde{f}(0)| = \operatorname{Re} \ln \tilde{f}(0) = \frac{1}{2\pi} \int_{-\pi}^{\pi} \ln|\tilde{f}(e^{i\theta})|\, d\theta$$

$$+ \frac{1}{\pi} \int_{\beta}^{1} \frac{e(x+i0)}{x}\, dx$$

$$\leq \frac{1}{2\pi} \int_{-\pi}^{\pi} \ln|\tilde{f}(e^{i\theta})|\, d\theta + \int_{\beta}^{1} \frac{dx}{x}$$

and

XIII (24) $$|f(0)| = |\tilde{f}(0)| \leq \frac{1}{\beta} \exp \frac{1}{2\pi} \int_{-\pi}^{\pi} \ln|\tilde{f}(e^{i\theta})|\, d\theta$$

On the other hand

$$\frac{1}{2\pi} \int_{-\pi}^{\pi} \ln|\tilde{f}(e^{i\theta})|\, d\theta$$

$$= \frac{1}{2\pi} \int_{-\pi}^{\pi} \ln|f(e^{i\theta})|\, d\theta - \sum_{j=1}^{n} \frac{1}{2\pi} \int_{-\pi}^{\pi} \ln\left|\frac{e^{i\theta} - r_j e^{i\theta_j}}{- r_j e^{i\theta_j}}\right| =$$

$$= \frac{1}{2\pi} \int_{-\pi}^{\pi} \ln |f(e^{i\theta})| \, d\theta + \sum_{j=1}^{n} \ln r_j$$

so that

$$\text{XIII (25)} \quad \exp \frac{1}{2\pi} \int_{-\pi}^{\pi} \ln |\hat{f}(e^{i\theta})| \, d\theta \leq \prod_{j=1}^{n} r_j \exp \frac{1}{2\pi} \int_{-\pi}^{\pi} \ln |f(e^{i\theta})| \, d\theta .$$

Combining XIII (24) and XIII (25) we have proved the lemma since $r_j < 1$.
Combining XIII (21) and XIII (23) we also get an estimate of the form

$$\text{XIII (26)} \quad \digamma(s_1, t, u_1) - \digamma(4-t, t, 0) \geq H_2(\digamma(s_1, t, u_1), s_1, t).$$

If we take $t = 2, \ s_1 = 3$, numerical calculations give

$$|\digamma(2,2,0)| \leq 19 \ ; \ |\digamma(3,2,-1)| \leq 75$$

and also for the symmetrical point

$$-50 \leq \digamma\left(\frac{4}{3}, \frac{4}{3}, \frac{4}{3}\right) \leq 8 .$$

This corresponds to the following estimate for the Chew - Mandelstam parameter λ :

$$-2.6 \leq \lambda \leq 17 .$$

These results allow us to get an absolute sum rule of the form VII(6) on the total cross-section. Indeed we have

$$0 < \frac{6}{\pi} \int_{4}^{\infty} \frac{A_s(s',2)(s'-1)}{s'(s'-3)(s'+1)(s'-2)} ds' = \digamma(3,2,-1) - \digamma(2,2,0) < 94$$

which by inequality VII(5) gives a nonlinear sum rule on the total cross-section. Also these estimates permit to obtain a lower bound on $\frac{d}{dk} \delta_0(k)$. This in turn gives a lower-bound for the $\pi^0 \pi^0$- scattering length [B 7]:

$$a_0^{\pi^0 \pi^0} = \frac{1}{3} a_0^{I=0} + \frac{2}{3} a_0^{I=2} > -4 \mu^{-1} .$$

The πN analysis yields a value $\cong 0.4\,\mu^{-1}$, assuming that $a_0^{I=2} = 0$ [D 1]. From current algebra and the hypothesis of partially conserved axial vector-current, S. Weinberg found a value $\cong 0.022\,\mu^{-1}$ [W 2], so conversely one may conclude that the above result forbids certain extrapolations of the off-shell amplitude obtained from current algebra.

Now the s-wave scattering length is determined by the s-wave phase shift through

$$a_0 = \lim_{k \to 0} \frac{f_0(k)}{k} = \lim_{k \to 0} \frac{\exp i\,\delta_0(k) \cdot \sin \delta_0(k)}{k}$$

$$= \lim_{k \to 0} \frac{1}{k \operatorname{ctg} \delta_0(k)} .$$

If we expand $\delta_0(k)$ around $k = 0$ and use $\delta_0(0) = n\pi$ we obtain

$$a_0 = \frac{d\,\delta_0}{dk}(0) .$$

The quantity a_0 has the following meaning: Take a non-relativistic scattering of two particles, where the interaction vanishes beyond a distance a_0, then [W 3]

$$\frac{d\,\delta_0(k)}{dk} > -a_0 + \frac{1}{2k} \sin 2(\delta_0 + a_0 k)$$

In order to obtain a lower bound for a_0 in the $\pi^0\pi^0$ - scattering case, it is therefore necessary to obtain a lower bound on $\frac{d}{dk}\delta_0(k)$. Bonnier and Vinh Mau proceeded as follows:

First write

$$\text{XIII (27)} \qquad S_l(k) = e^{2i\delta_l(k)} = 1 + 2i\, f_l(k)$$

$$= 1 + \frac{8i}{\sqrt{s(s-4)}} \int_0^{\frac{1}{2}(4-s)} F(s,t)\, P_l\left(1 + \frac{2t}{s-4}\right) dt .$$

Thus we need bounds on $|F(s,t)|$ for complex s and t. Write

$$F(s,t) = F(s,t) - F(s_0,t) + F(s_0,t) - F(s_0,t_0) + F(s_0,t_0)$$

and take $s_0 = t_0 = 2$. Then we have the bound $|F(2,2)| < 19$. Also

XIII (28)

$$\Phi(t,2) - \Phi(2,2) = \frac{2(2-t)t}{4} \int_4^\infty \frac{A_s(s',2)(s'-1)\,ds'}{s'(s'-2)(s'-t)(s'-2+t)} .$$

The problem is to find an estimate for the right hand side. Now

$$\Phi(3,2) - \Phi(2,2) = \frac{6}{\pi} \int_4^\infty \frac{A_s(s',2)(s'-1)\,ds'}{(s'-3)(s'+1)(s'-2)\,s'}$$

is positive and $|\Phi(3,2)| < 75$; $|\Phi(2,2)| < 19$

so

$$0 < \frac{6}{\pi} \int_4^\infty \frac{A_s(s',2)(s'-1)}{(s'-3)(s'+1)(s'-2)s'}\,ds' < 94 .$$

If we insert this in XIII (28), we obtain

$$\text{XIII (29)} \quad |\Phi(t,2) - \Phi(2,2)| < \frac{94}{3}\,|t(t-2)| \sup_{4 \le s'} \frac{(s'+1)(s'-3)}{|s'-t|\,|s'-2+t|} .$$

To obtain an upper bound for $|\Phi(s,t) - \Phi(2,t)|$ we first note that for $|t|<4$

$$|A_s(s',t)| \le A_s(s',|t|)$$

Also $\quad \Phi(4-|t|,|t|) = \Phi(|t|,0) \le \Phi(4,0) < 0$

and $\quad \Phi(\tfrac{1}{2}(4-|t|),|t|) \ge \Phi(\tfrac{4}{3},\tfrac{4}{3}) \ge -50$

(see the discussion of XIII (14)) , so a trick similar to the one we just used gives

$$|\Phi(s,t) - \Phi(2,t)| \le 50\, \frac{|s-2|\,|s-2+t|}{(4-|t|)^2} \times$$

$$\times \sup_{4 \le s'} \frac{|(2s'+t-4)(2s'+|t|-4)\,s'(s'-4+|t|)|}{|(s'-s)(s'-u)(s'-2+t)(s'-2)|}$$

Combining these estimates, we obtain a bound for $\Psi'(s,t)$ $(|t|<4\,;\ s,u \notin [4,\infty))$, Obviously this gives a bound for $S_\ell(k)$ if $|s-4|<8$, s not being on the cuts.

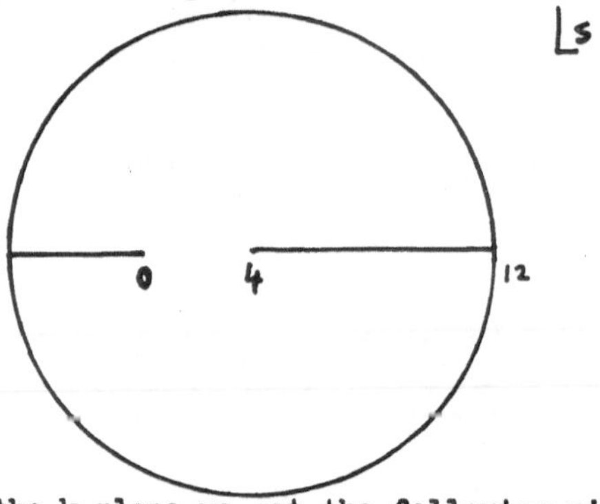

Transforming this to the k-plane, we get the following picture :

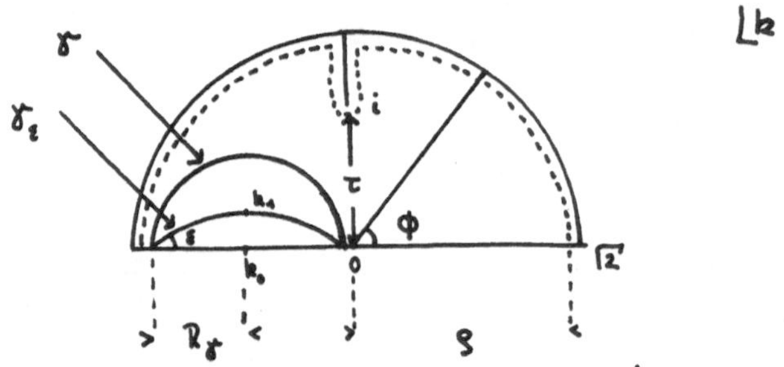

On the real axis we have $|S_\ell(k)| \leq 1$ and on γ we have an estimate of the form $|S_0(k)| \leq B_\gamma$ and hence on γ_ε

$$|S_0(k)| < \exp \frac{2\varepsilon}{\pi} \log B_\gamma$$

(see page 52) .

If we consider the point k_1 , then $\mathrm{Re}\, k_0 = \mathrm{Re}\, k_1,\ \mathrm{Im}\, k_1 \cong R_\gamma \cdot \frac{\varepsilon}{2}$ so

$$|S_0(k_1)| \leq \exp \frac{4}{\pi} \frac{\mathrm{Im}\, k_1}{R_\gamma} \log B_\gamma .$$

On the other hand

$$S_0(k_1) \cong S_0(k_0) + (k_1-k_0)\, 2i \frac{d}{dk}\delta_0(k_0) \cdot S_0(k_0)$$

$$S_0(k_1) \cong S_0(k_0) + (k_1 - k_0)\, 2i \frac{d}{dk} \delta_0(k_0) \cdot S_0(k_0)$$

so since $|S_0(k_0)| = 1$,

$$|S_0(k_1)| \cong 1 - 2 \, \mathrm{Im}\, k_1 \frac{d\delta_0}{dk}(k_0).$$

Therefore

$$\frac{d\delta_0}{dk}(k_0) > - \frac{2}{\pi R_\gamma} \log B_\gamma .$$

If we optimize R_γ and B_γ, we get a lower bound for $\frac{d\delta_0}{dk}(k_0)$ which will be a function of k_0.

The discussion we gave was just a qualitative estimate. In a more refined discussion, one takes into account the positivity of the discontinuity of f_0 across the relevant portion of the left hand cut and one obtains a kind of Poisson formula by arguments similar to those which led to the lemma of page 99.

We just quote the result for $k_0 = 0$:

$$\frac{d\delta_0}{dk} > - \frac{1}{\pi} \int_0^\pi \frac{\ln |S_0(\rho e^{i\phi})|}{\rho} \sin\phi \, d\phi - \frac{1}{2}\left(\frac{1}{\tau} - \frac{1}{\rho}\right).$$

By this the original estimate $a_0 > -4\mu^{-1}$ obtained by Bonnier and Vinh Mau could be improved to $a_0 > -3.5\mu^{-1}$.

At last we want to discuss another application of estimates on $|F(s,t,u)|$ [M 14]. We leave it to the reader to decide whether the following may be only of academic interest or not.

Assume we are given a scattering of stable neutral scalar particles σ of mass m through a Lagrangian of the form

$$\lambda : \phi^3 : + \hat{\lambda} : \phi^4 :$$

The fourth order term is introduced in order to "guarantee" the positivity of the energy spectrum. But then graphs of the following form

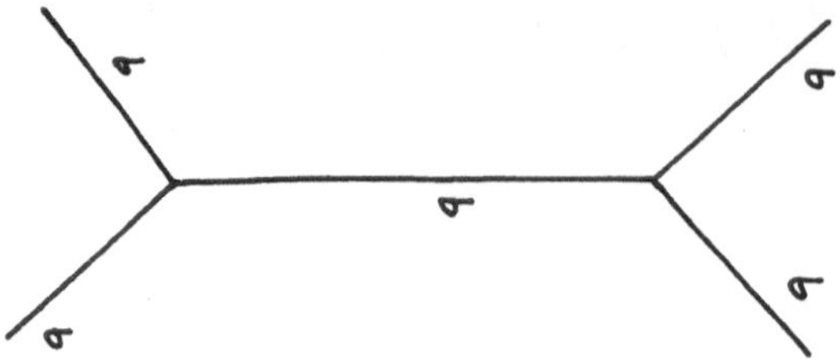

appear, which will modify the scattering amplitude to

$$\Phi(s,t,u) = -g^2\left(\frac{1}{s-m^2}+\frac{1}{t-m^2}+\frac{1}{u-m^2}\right)+C(t)$$
$$+\frac{1}{\pi}\int_{4m^2}^{\infty}\frac{A_s(s',t)}{s'^2}\left\{\frac{s^2}{s'-s}+\frac{u^2}{s'-u}\right\}ds'$$

where $\frac{\lambda^2}{4\pi}=\frac{1}{36}\frac{g^2}{4\pi}$. Apart from the pole terms, we assume $\Phi(s,t,u)$ to have the same analyticity properties as those we obtained in the preceeding sections. In order to get estimates, we will combine XIII (1) with the estimate $|\Phi(s,\cos\theta)|\leq\bar{\Phi}(s)$ for physical $\cos\theta$:

XIII (30) $$A_s(s,t)\geq\phi(s,t)\,|\Phi(s,4m^2-\bar{s}-t,\bar{s}+t-s)|^2$$
$$s\geq\bar{s}+t>4m^2\,;\quad 0<t<4m^2 .$$

Defining the function

$$\tilde{\Phi}(s,t,u) = (s-m^2)(t-m^2)(u-m^2)\,\Phi(s,t,u)$$

we conclude that for fixed $t_0\leq t'<0$, it is analytic in s with cuts $-\infty<s\leq-t'$ and $4m^2\leq s<\infty$. Also it is easy to see that the discontinuity of $\tilde{\Phi}(s',4m^2-\bar{s}-t,\bar{s}+t-s')$ across the unphysical cut $4m^2<s'<\bar{s}+t$ is positive if $4m^2<\bar{s}+t<5m^2$ (use crossing and V (14)) .

In our "inverse bootstrap technique" we may therefore use XIII (30) twice in the dispersion relation for $\Phi(s_1,t,u_1)-\Phi(4m^2-t,t,0)$ and then Poisson's inequality (resp. XIII (23) with $\bar{s}=s_1$) for $\tilde{\Phi}$ and obtain estimates of the form

XIII (31) $$\Phi(s_1,t,u_1)-\Phi(4m^2-t,t,0)\geq g^2\sigma+c_1|\Phi(s_1,t,u_1)|^2$$
$$\Phi(s_1,t,u_1)-\Phi(4m^2-t,t,0)\geq g^2\sigma+c_1|\Phi(4m^2-t,t,0)|^2$$

where $s_1 = 4m^2 - t + \tau \quad , 0 < \tau < m^2 \quad , 2m^2 < t < 3m^2$.

σ, c_1, c_2 are positive and may explicitly be calculated. Equation XIII (31) combined with

$$|F(s_1,t,u_1)| + |F(4m^2-t,t,0)| \geq F(s_1,t,u_1) - F(4m^2-t,t,0)$$

gives
$$g^2 < \max\left\{\frac{1}{\sigma c_1}, \frac{1}{\sigma c_2}\right\}.$$

A rough numerical estimate gives

$$\frac{g^2}{4\pi} < 1.5 \cdot 10^6 \, m^2$$

which is a large number, but of course this calculation was only carried through to show that there <u>exists</u> an upper bound.

XIV Particles with spin, superconvergence relations

In this chapter we will discuss the scattering of two particles with spin. Up till now all our results were derived for spinless particles. We will show that the introduction of spin does not change the results in any essential way. Remember that the starting point was to show analyticity of $F(s,t)$ for $|t| < R$ and for s in the cut plane. To prove this we have used the positivity properties of the absorptive part.

The simplest spin-dependent case, pion-nucleon scattering, has first been discussed by Sommer [55]: Let the 4×4 matrix W be the Dirac spinor scattering amplitude. Then we may write

$$W^{\pm} = -A^{\pm}(s,t,u) + \frac{1}{2}(\not{q}_1 + \not{q}_2)\, B^{\pm}(s,t,u)$$

where q_1 and q_2 are the momenta of the incoming and outgoing pion and "$\pm$" give the isospin combinations

$$\frac{1}{2}\left(T_{\pi^- p} \pm T_{\pi^+ p}\right).$$

If we choose

$$F^{\pm}(s,t,u) = A^{\pm}(s,t,u) + \frac{s-u}{4M} B^{\pm}(s,t,u)$$

then these amplitudes have the right positivity properties in both the s- and u-channel. For $t = 0$ they are just the spin-nonflip amplitudes.

The general case has been discussed by G. Mahoux and A. Martin [M 2], G. Mahoux [M 1] and J.S. Bell [B 2] . In our discussion we will follow the arguments of J.S. Bell. First of all the results of Lehmann [L 1] on analyticity in the Lehmann ellipses, those of Bros, Epstein and Glaser on analyticity in s and t (see page 17) and on crossing (see page 18) still hold for scattering amplitudes of particles with spin, provided they have no kinematical singularities.

Therefore all we have to do is to find amplitudes which are free of kinematical singularities and have the positivity property in the s- and u - channel.

We start again from the scattering of two particles A and B , A having spin $\frac{1}{2}$, B spin zero. Let p_1 and p_2 be the initial and final four-momenta of A , q_1 and q_2 those of B . Again let also W be the 4×4 - spinor scattering amplitude, which is supposed to be free of kinematical singularities. If a_1 and a_2 specify initial and final spin states of A , then the center-of-mass amplitudes with the relative three-momenta $\underset{\sim}{k}_1, \underset{\sim}{k}_2$

$$< \underset{\sim}{k}_2 , a_2 | T | \underset{\sim}{k}_1 , a_1 >$$

are related to W by

$$< \underset{\sim}{k}_2 , a_2 | T | \underset{\sim}{k}_1 , a_1 > = N \, \bar{u}_2(a_2) \, W \, u_1(a_1) \, .$$

$u_1(a_1)$ and $u_2(a_2)$ are Dirac spinors for the momenta p_1 and p_2 and spin projections a_1 and a_2 . Of course p_1, q_1 and p_2, q_2 are specialized to their center-of-mass values

$$p_i = \left(\sqrt{m_A^2 + \underset{\sim}{k}_i^2}, \underset{\sim}{k}_i\right) \qquad i = 1, 2$$

$$q_i = \left(\sqrt{m_B^2 + \underset{\sim}{k}_i^2}, -\underset{\sim}{k}_i\right)$$

N is a real positive normalization factor which depends only on s . The part of W which has physical relevance is obtained by projecting onto the physical spin states:

$$\left(\frac{1}{2 m_A}\right)^2 \sum_{a_2} u_2(a_2) \, \bar{u}_2(a_2) \, W \sum_{a_1} u_1(a_1) \, \bar{u}_1(a_1)$$

$$= \Lambda^+(p_2) \, W \, \Lambda^+(p_1)$$

$$\Lambda^+(p_i) = \frac{\not{p}_i + m_A}{2 m_A} \, .$$

The trace of this would give an invariant amplitude, but it is found not to have the desired positivity property. Essentially the only other invariant with simple crossing properties is

$$\mathbb{F}(s,t,u) \underset{def}{=} -\operatorname{Trace}(\not q_1 + \not q_2)\,\Lambda^+(p_2)\,\mathbb{W}\,\Lambda^+(p_1)$$

such that

$$(2m_A)^2 N\,\mathbb{F}(s,t,u) = -\sum_{a_1,a_2} \langle \underset{\sim}{k}_2, a_2 | \mathbb{T} | \underset{\sim}{k}_1, a_1 \rangle\, \bar{u}_1(a_1)(\not q_1 + \not q_2)\, u_2(a_2)$$

which gives

$$(2m_A)^2 N\,\operatorname{Im}\mathbb{F}(s,t,u) = -\sum_{a_1,a_2} \langle \underset{\sim}{k}_2, a_2 | \mathbb{T}^+\mathbb{T} | \underset{\sim}{k}_1, a_1 \rangle \times$$

$$\times\, \bar{u}_1(a_1)(\not q_1 + \not q_2)\, u_2(a_2),$$

if we use the unitarity condition

$$\frac{1}{2i}(\mathbb{T} - \mathbb{T}^+) = \mathbb{T}^+\mathbb{T}.$$

Suppose now that the axis of spin quantization is in the direction of $\underset{\sim}{k}_1 \times \underset{\sim}{k}_2$. Then by explicit calculation

XIV (1) $$-\bar{u}_1(a_1)(\not q_1 + \not q_2)\, u_2(a_2) = \delta_{a_1,a_2}\left(X + Y e^{2ia_1\theta}\right)$$

where Θ is the scattering angle and

XIV (2) $$X = 2(p_0 + m_A)(q_0 + p_0 - m_A) \geq 0$$
$$Y = 2(p_0 - m_A)(q_0 + p_0 + m_A) \geq 0$$
$$p_0 = (m_A^2 + k_1^2)^{\frac{1}{2}}\ ; \quad q_0 = (m_A^2 + k_1^2)^{\frac{1}{2}}.$$

We have used the phase convention

$$|k_2, a\rangle = e^{-i\theta(J-a)}\,|k_1, a\rangle$$

where J is the angular momentum operator along the $\underset{\sim}{k}_1 \times \underset{\sim}{k}_2$ - direction. If we repeat our discussion on page 7 we get

XIV (3) $$\langle \underset{\sim}{k}_2, a | \mathbb{T}^+\mathbb{T} | \underset{\sim}{k}_1, a \rangle = \sum_n C_n^a\, e^{in\theta}$$

with $C_n^a \geq 0$ and due to the phase convention n runs through all integers. Thus we have

$$\text{XIV (4)} \qquad (2m_A)^2 N \,\mathrm{Im}\, F(s,t,u) = \sum_{n=-\infty}^{+\infty} C'_n(s)\, e^{in\theta}$$

$$C'_n(s) = \sum_a \{ X(s)\, C_n^a(s) + Y(s)\, C_{n-2a}^a(s) \}$$

Since $\mathrm{Im}\, F$ is real

$$(2m_A)^2 N \,\mathrm{Im}\, F(s,t,u) = C'_0(s) + \sum_1^\infty C'_n(s) \cos n\theta$$

so $\mathrm{Im}\, F(s,t,u)$ is a function of positive type in θ. Because of the factor $(\not{q}_1 + \not{q}_2)$, $F(s,t,u)$ has the following crossing property

$$F(s,t,u) = - F(u,t,s)$$

and therefore

$$F'(s,t,u) = (s-u)\, F(s,t,u)$$

satisfies

$$F'(s,t,u) = F'(u,t,s),$$

and it is easy to see that this function has the right positivity properties in both channels. Thus we may continue $F'(s,t,u)$ analytically to positive t at least as far as to some $t = R$, fixed independently of s. We will show that this information on this particular amplitude suffices to derive the Froissart bound for the unpolarized differential cross-section:

$$\frac{d\sigma}{d\Omega} = \frac{4}{2s} \sum_{a_1} \sum_{a_2} |\langle \underset{\sim}{k}_2, a_2 | T | \underset{\sim}{k}_1, a_1 \rangle|^2$$

The factor $\frac{1}{2}$ arises because we are averaging over the initial spin states. Consider in particular the spin averaged amplitude

$$\text{XIV (5)} \qquad f(s, \cos\theta) = \frac{1}{2} \sum_a \langle k_2, a | T | k_1, a \rangle .$$

With the spin averaged over, orbital angular momentum is conserved, so $f(s,\cos\theta)$ has a similar partial wave expansion as the scattering amplitude for spinless particles.

$$f(s,\cos\theta) = \frac{s^{\frac{1}{2}}}{2k} \sum_{\ell=0}^{\infty} (2\ell+1)\, f_\ell(s)\, P_\ell(\cos\theta),$$

$$k = |\underset{\sim}{k}_1| = |\underset{\sim}{k}_2|.$$

Unitarity again gives

$$|f_\ell(s)|^2 \leq \operatorname{Im} f_\ell(s) \leq 1.$$

This observation is due to Yamamoto [Y 1].

Now

$$\sum_a C_0^a + \sum_{n=1}^{\infty} \Big(\sum_a C_n^a \Big) \cos n\theta$$

is essentially the series for $\operatorname{Im} f(s,\cos\theta)$. Since

$$X^{-1} C_n^1 \geq \sum_a C_n^a \geq 0$$

it follows that also this series converges for $\cos\theta < 1 + \frac{R}{2k^2}$. This again gives the Froissart bound

$$|f(s,\cos\theta)| \leq C s (\log s)^2$$

$$\frac{d\sigma}{d\Omega} \leq C \cdot (\log s)^2.$$

In order to generalize this result for arbitrary spin, the Wigner-Bargmann formalism [B 1] is convenient. The wave function of a particle of spin $\frac{1}{2}\varrho$ transforms like the symmetrical part of a product of ϱ Dirac wave functions. The generalized spinor amplitude will then have ϱ paris of Dirac indices for particle A of spin $\frac{1}{2}\varrho$ and σ pairs for particle B of spin $\frac{1}{2}\sigma$:

$$W_{\{\ell_2^i,\, \ell_1^i\},\, \{m_2^j,\, m_1^j\}} \qquad \begin{array}{l} (1 \leq i \leq \varrho) \\ (1 \leq j \leq \sigma) \end{array}$$

and W can be supposed to be symmetric in each of the set of variables $\{\ell_2^i\} \ldots \{m_1^j\}$.

Each pair of indices can now be treated separately as before. Thus the scattering matrix elements

$$\langle \underset{\sim}{k}_2, a_2^1 \ldots a_2^\varrho; b_2^1 \ldots b_2^\sigma | T | \underset{\sim}{k}_1, a_1^1 \ldots a_1^\varrho; b_1^1 \ldots b_1^\sigma \rangle$$

are obtained (apart from a normalization factor) by saturating the indices of W with wave functions

$$\prod_{i=1}^{\varrho} \bar{u}_{\ell_2^i}(a_2^i)\, u_{\ell_1^i}(a_1^i) \prod_{j=1}^{\sigma} \bar{u}_{m_2^j}(b_2^j)\, u_{m_1^j}(b_1^j)$$

The invariant amplitude $F(s,t,u)$ is found by saturating the indices of W with factors

$$\prod_{i=1}^{\varrho} \left(\Lambda^+(p_1)(\not{q}_1+\not{q}_2)\Lambda^+(p_2)\right)_{\ell_2^i \ell_1^i} \prod_{j=1}^{\sigma} \left(\Lambda^+(q_1)(\not{p}_1+\not{p}_2)\Lambda^+(q_2)\right)_{m_2^j m_1^j} .$$

Thus the problem essentially factorizes. Proceeding as before we get a representation with positive C's :

$$\operatorname{Im} F(s,t,u) = \sum_n e^{in\theta} \sum_{a,b} C_n^{a^1 \dots a^\varrho,\, b^1 \dots b^\sigma} \times$$

$$\times \prod_{\ell=1}^{\varrho} (X_A + Y_A\, e^{2ia^\ell\theta}) \prod_{j=1}^{\sigma} (X_A + Y_B\, e^{2ib^j\theta})$$

where the coefficients X_A and Y_A are given by XIV (2) and X_B and Y_B are analogous expressions in which the roles of the two particles are inter-changed. Thus we once again get a representation

$$\operatorname{Im} F(s,t,u) = \sum C_n \cos n\theta$$

with positive coefficients. In the crossing relation a factor (-1) arises for each pair of Dirac indices. Therefore

$$F'(s,t,u) = (s-u)^{\zeta}\, F(s,t,u)$$

($\zeta = 0$ if $\sigma+\varrho$ even, $\zeta = 1$ if $\sigma+\varrho$ odd) has the right positivity property in both channels. This will give analyticity and the Froissart bound for the spin averaged amplitude defined in analogy with XIV (5) and for the unpolarized elastic differential cross section.

In this derivation we have again assumed the spinor amplitude W to be free of kinematical singularities. Instead of working in the Wigner-Bargmann formalism, one could alternatively use wave functions which transform as $(j,0)$ or $(0,j)$. Therefore it does not matter in which version the sesquilinear form giving the scattering amplitude is supposed to have coefficients free from kinematical singularities. This kinematical equivalence holds for any pair of wave functions representing a massive particle of given spin, as has been proved by F. Yndurain [B 2].

In order to bound amplitudes other than the spin average, a more general partial wave expansion is required.

We first quote a <u>theorem</u> [M 1] :

Suppose we are given a scattering amplitude of two stable massive particles A with spin $\frac{1}{2}\sigma$ and B with spin $\frac{1}{2}\rho$. Assume that this amplitude is free of kinematical singularities and that it satisfies a fixed t dispersion relation $(t_0 \leq t \leq 0)$. Then this amplitude also satisfies fixed t dispersion relations for $|t| < R$ and some $R > 0$. The number of subtractions increases at most by one.

<u>Remark:</u> Actually it is not necessary to assume validity of fixed t dispersion relations (see p. 30). For simplicity, however, we will restrict ourselves to that case.
Now consider the partial wave expansion of the helicity amplitude [J 1]:

XIV (6) $$M_{\lambda_A,\lambda_B;\lambda'_A,\lambda'_B}(s,\cos\theta_s) = \sum_J (2J+1)\, f^J_{\lambda_A,\lambda_B;\lambda'_A\lambda'_B}(s)\, d^J_{\lambda'\lambda}(\theta_s)$$

where $$\lambda = \lambda_A - \lambda_B\,; \quad \lambda' = \lambda'_A - \lambda'_B$$

and $$d^J_{\lambda'\lambda}(\theta) = \langle J^2_{op.} = J(J+1), J_z = \lambda' |\, \exp - i J_y \cdot \theta \,| J^2_{op} = J(J+1), J_z = \lambda \rangle .$$

Apart from the dynamical singularities, the amplitudes $M_{\lambda_A,\lambda_B;\lambda'_A,\lambda'_B}$ and $f^J_{\lambda_A,\lambda_B;\lambda'_A,\lambda'_B}$ have kinematical singularities. Those of the partial wave amplitudes are at the threshold $s = (M_A + M_B)^2$, at the pseudothreshold $s = (M_A - M_B)^2$ and at the point $s = 0$. $d^J_{\lambda'\lambda}(\theta)$ has only kinematical singularities. For a general discussion of the kinematical singularities see [C 3], [J 3], [T 3], [W 1] and [W 4]. In spite of these kinematical singularities the absorptive part can be made well defined and the following unitarity relation holds:

XIV (7) $$\sum_{\lambda'_A,\lambda'_B} |\, f^J_{\lambda_A,\lambda_B;\lambda'_A,\lambda'_B}(s)|^2 \leq \frac{s^{\frac{1}{2}}}{2k}\, \mathrm{Im}\, f^J_{\lambda_A,\lambda_B;\lambda_A,\lambda_B}(s) \leq \frac{s}{4k^2} .$$

Now, from the explicit form of $d^J_{\lambda'\lambda}(\theta)$ the following may be shown:
If $\rho + \sigma$ is even, then

XIV (8) $$M_{\lambda_A,\lambda_B;\lambda_A,\lambda_B}(s,t)$$

has no kinematical singularities in t .

If $\rho + \sigma$ is odd, then

$$\text{XIV (9)} \qquad \cos\frac{\theta}{2}\, M_{\lambda_A,\lambda_B;\lambda_A,\lambda_B}(s,t)$$

has no kinematical singularities in t. Therefore the above theorem is applicable to XIV (8) (resp. XIV (9)), and we get for physical s an analytic function in $|t| < R$. This remains true if we take the absorptive parts

$$\text{XIV (10)} \qquad \mathrm{Abs}\, M_{\lambda_A,\lambda_B;\lambda_A,\lambda_B}(s,t) = \sum_J (2J+1)\, \mathrm{Im}\, f^J_{\lambda_A,\lambda_B;\lambda_A,\lambda_B}(s)\, d^J_{\lambda\lambda}(\theta)$$

[resp.

$$\text{XIV (11)} \qquad \cos\frac{\theta}{2}\, \mathrm{Abs}\, M_{\lambda_A,\lambda_B;\lambda_A\lambda_B}(s,t) = \sum_J (2J+1)\, \mathrm{Im}\, f^J_{\lambda_A,\lambda_B;\lambda_A\lambda_B}(s) \cos\frac{\theta}{2}\, d^J_{\lambda\lambda}(\theta)].$$

Equation XIV (7) shows that $\mathrm{Im}\, f_{\lambda_A,\lambda_B;\lambda_A,\lambda_B}(s)$ is positive. Therefore we may repeat the argument on page 36 to show that XIV (10) (resp. XIV (11)) is analytic in $\cos\theta$ in an ellipse with foci ± 1 and semimajor axis $\cos\Theta_0(s) = 1 + \frac{R}{2k^2}$. This is so because the $d^J_{\lambda\lambda}(\theta)$ (resp. $\cos\frac{\theta}{2}\, d^J_{\lambda\lambda}(\theta)$) have properties similar to $P_\ell(\cos\theta)$; in particular they are of positive type (see Appendix). Therefore we get the estimates

$$\frac{2k}{s^{\frac{1}{2}}}\, |f_{\lambda_A,\lambda_B;\lambda'_A\lambda'_B}(s)|^2 \leq \mathrm{Im}\, f_{\lambda_A,\lambda_B;\lambda_A,\lambda_B}(s)$$

$$\leq \frac{\mathrm{Abs}\, M_{\lambda_A,\lambda_B;\lambda_A,\lambda_B}(s,t=R)}{d^J_{\lambda\lambda}(i\hat{\theta})}; \qquad \hat{\theta} = \cosh^{-1}\left(1+\frac{R}{2k^2}\right).$$

Now

$$d^J_{\lambda\lambda}(i\hat{\theta}) \geq \frac{\phi_0}{\pi}\left(\cosh\hat{\theta} + \cos\phi_0 \sinh\hat{\theta}\right)^{J-|\lambda|} \left(\cosh\frac{\hat{\theta}}{2}\right)^{|\lambda|}.$$

If we use the estimate

$$\mathrm{Abs}\, M_{\lambda_A,\lambda_B;\lambda_A\lambda_B}(s,t=R) < s^N$$

then we only have contributions in the sum

XIV (6) up to $L(s) \approx \mathrm{const}\; s^{\frac{1}{2}} \cdot \log s$. Therefore asymptotically

$$\text{XIV (12)} \qquad |M_{\lambda_A,\lambda_B;\lambda'_A,\lambda'_B}(s,t)| \leq \mathrm{const}\; s (\log s)^2 \quad \text{for } t \leq 0,\ s \to +\infty,$$

which again is the Froissart bound.

With the help of these bounds we are now able to prove the existence of superconvergence relations, which are sum rules involving only the absorptive part of an amplitude [M 1]: Consider as an example

$$\mathcal{F}(s,t) = \frac{1}{\pi} \int_{s_0}^{\infty} \frac{A_s(s',t)}{s'-s}\, ds'$$

and assume $s\mathcal{F}(s,t)$ goes to zero for $s \to \infty$ and fixed t. Then we get

$$\int_{s_0}^{\infty} A_s(s',t)\, ds' = 0$$

Relations of this kind were first derived in the framework of current algebra. Since in these relations only on-shell amplitudes are involved, it is reasonable to expect that they might be derived more directly. (We shall follow the method of L. Trueman [T 4] and R. Odorico [O 1]).

Let us start with the helicity amplitude

$M^t_{\lambda_B,\lambda_{\bar{B}};\lambda_A,\lambda_{\bar{A}}}$ in the t-channel, i.e. with the amplitude describing the process $A+\bar{A} \to B+\bar{B}$. With Comonave [C 3] let us define the partially regularized amplitudes

$$\hat{M}^t_{\lambda_B,\lambda_{\bar{B}};\lambda_A,\lambda_{\bar{A}}} = \frac{M^t_{\lambda_B,\lambda_{\bar{B}};\lambda_A,\lambda_{\bar{A}}}}{\left(\sin\frac{\Theta_t}{2}\right)^{|\lambda'-\lambda|}\left(\cos\frac{\Theta_t}{2}\right)^{|\lambda+\lambda'|}}$$

where $\lambda' = \lambda_A - \lambda_{\bar{A}}$; $\lambda = \lambda_B - \lambda_{\bar{B}}$ and where Θ_t, the scattering angle in the center-of-mass system in the t-channel, is given by

$$\text{XIV (13)} \qquad \cos\Theta_t = \frac{s-u}{\sqrt{(t-4m_A^2)(t-4M_B^2)}}$$

$$\sin\Theta_t = 2\,\frac{\sqrt{su-(m_A^2-m_B^2)^2}}{\sqrt{(t-4m_A^2)(t-4M_B^2)}}$$

The amplitudes $\hat{M}^t$ have for fixed t no kinematical singularities in s. According to Comonave

XIV (14)

$$\hat{M}^t_{\lambda_B,\lambda_{\bar{B}};\lambda_A,\lambda_{\bar{A}}} \quad , \text{ if } \lambda+\lambda' \text{ is even ,}$$

$$\frac{1}{\sqrt{-t}}\,\hat{M}^t_{\lambda_B,\lambda_{\bar{B}};\lambda_A,\lambda_{\bar{A}}} \quad , \text{ if } \lambda+\lambda' \text{ is odd}$$

have no kinematical singularities around $t = 0$, although they generally might have singularities at the thresholds $t = 4m_A^2$ and $t = 4m_B^2$. In any case, these singularities are outside the circle $|t| < R$. Therefore the amplitudes satisfy dispersion relations in s for fixed $|t| < R$.

Let us look at the asymptotic behaviour in s of these amplitudes when t is real and negative. For simplicity we will only discuss the case where $\lambda + \lambda'$ is even. The helicity amplitudes M^t in the t-channel may be expressed in terms of the helicity amplitudes M^s in the s-channel by means of a crossing matrix. All we need to know about this matrix is that it may be expressed in terms of rotation matrices, whose angles are real in the physical region of the s-channel. Therefore all elements of the crossing matrix are bounded by 1 in norm if s and t have physical values, so the amplitudes M^t satisfy Froissart bounds. According to XIV (13) $\cos\frac{\theta_t}{2}$ and $\sin\frac{\theta_t}{2}$ behave like $\sqrt{s}$ for $s \to +\infty$ so we have for $\lambda + \lambda'$ even

XIV (15)
$$\left| \hat{M}^t_{\lambda_B, \lambda_{\bar{B}}; \lambda_A, \lambda_{\bar{A}}} \right| \leq C \cdot |s|^{1 - \mathrm{Max}(|\lambda|, |\lambda'|)} (\log|s|)^2$$
$$t \leq 0 \; ; \; s \to +\infty.$$

The same holds for $u \to +\infty$ (i.e. $s \to -\infty$). An application of the theorem of Phragmen-Lindelöf then shows XIV (15) holds for all complex directions in the cut plane and for $t \leq 0$.

If $\lambda + \lambda'$ is odd, the following asymptotic estimate for complex s and $t \leq 0$ may be obtained:

XIV (16)
$$\left| \frac{1}{\sqrt{-t}} M^t_{\lambda_B, \lambda_{\bar{B}}; \lambda_A, \lambda_{\bar{A}}} \right| < C \cdot |s|^{1 - \mathrm{Max}(|\lambda|, |\lambda'|)} (\log|s|)^3.$$

In particular if $N = \mathrm{Max}(|\lambda|, |\lambda'|) > 2$ the amplitudes XIV (14) are superconvergent, i.e. we have $N-2$ superconvergence relations of the form

XIV (17)
$$\begin{pmatrix} 1 \\ \text{or} \\ \frac{1}{\sqrt{-t}} \end{pmatrix} \times \left\{ \int_{(M_A+M_B)^2}^{\infty} ds \cdot s^n \cdot \mathrm{Abs}_s \, \hat{M}^t_{\lambda_B, \lambda_{\bar{B}}; \lambda_A, \lambda_{\bar{A}}}(s,t) - \int_{(M_A+M_B)^2}^{\infty} ds \cdot s^n \cdot \mathrm{Abs}_u \, \hat{M}^t_{\lambda_B, \lambda_{\bar{B}}; \lambda_A, \lambda_{\bar{A}}}(u,t) \right\} = 0,$$

$$t \leq 0 \; ; \; n = 0, 1 \ldots\ldots N-3.$$

This may happen, if one of the spins $\frac{1}{2}\sigma$ or $\frac{1}{2}\varsigma$ is at least equal to $\frac{3}{2}$
A somewhat academic example is the case of $\Omega - \pi$ scattering.

Without any additional dynamical information, nothing can be said about the scattering of two particles with spin 0 and 1 .

Appendix

In this appendix we collect some properties of the Legendre Polynomials.

Statements, which are quoted here but not proved, may be found in standard textbooks [E 4] [S 10] :

The Legendre Polynomials are solutions of the differential equation

$$(1 - x^2)\, P_\ell''(x) - 2x\, P_\ell'(x) + \ell \cdot (\ell+1)\, P_\ell(x) = 0 \tag{A 1}$$

with the boundary condition

$$P_\ell(\pm 1) = (\pm 1)^\ell . \tag{A 2}$$

They satisfy the following recurrence formula

$$(\ell+1)\, P_{\ell+1}(x) = (2\ell+1)\, x\, P_\ell(x) - \ell\, P_{\ell-1}(x) . \tag{A 3}$$

The following closed form holds:

$$P_\ell(\cos\Theta) = \sum_{k=0}^{\ell} g_k\, g_{\ell-k} \cos(\ell - 2k)\Theta \tag{A 4}$$

$$g_k = 2^{-2k} \binom{2k}{k}$$

which expresses P_ℓ in terms of Tchebichef polynomials.

Also

$$P_\ell(-x) = (-1)^\ell\, P_\ell(x) . \tag{A 5}$$

All zero's of P_ℓ lie in the interval [-1,1] . In this interval $|P_\ell(x)| \leq 1$.
Another estimate is

$$|P_\ell(\cos\Theta)| \leq 2\sqrt{\frac{1}{\pi(2\ell+1)\sin\Theta}} \quad ; \quad 0 < \Theta < \pi . \tag{A 6}$$

An integral representation is given by

(A 7) $$P_\ell(\cos\Theta) = \frac{1}{\pi}\int_0^\pi (\cos\Theta + i\sin\Theta\cos\phi)^\ell \, d\phi$$

and a generating function is given by

(A 8) $$(1 - 2xz + z^2)^{-\frac{1}{2}} = \sum_{\ell=0}^{\infty} P_\ell(x)\, z^\ell .$$

The Legendre Polynomials satisfy the following orthogonality relation in [-1,1]

(A 9) $$\int_{-1}^{+1} P_\ell(x)\, P_{\ell'}(x)\, dx = \frac{\delta_{\ell,\ell'}}{\ell + \frac{1}{2}} .$$

Complete induction, using A (3) , gives

(A 10) $$\sum_{\ell=0}^{L} (2\ell+1)\, P_\ell(x) = (L+1)\, \frac{P_{L+1}(x) - P_L(x)}{x-1} .$$

Similarly

(A 11) $$P_{\ell+1}(x) > P_\ell(x) \quad ; \quad x > 1$$

(A 12) $$\frac{P_{\ell+1}(x)}{P_{\ell+1}(x')} < \frac{P_\ell(x)}{P_\ell(x')} \quad ; \quad 1 \leq x < x' .$$

Put $x = \cosh\Psi$ for $x > 1$. Then (A 4) gives

$$P_\ell(\cosh\Psi) \geq 2^{-2\ell}\binom{2\ell}{\ell}\cosh\ell\Psi$$
$$\geq 2^{-(2\ell+1)}\binom{2\ell}{\ell}\, e^{\ell\Psi} = 2^{-(2\ell+1)}\binom{2\ell}{\ell}\left(x+\sqrt{x^2-1}\right)^\ell .$$

Therefore

(A 13) $$P_\ell(x) \geq \frac{c_0}{\sqrt{2\ell+1}}\left(x+\sqrt{x^2-1}\right)^\ell \geq \frac{c_0}{\sqrt{2\ell+1}}\left(1+\sqrt{2(x-1)}\right)^\ell$$
$$x \geq 1 ,$$

where Stirling's formula and the trivial estimate

$$\left(x + \sqrt{x^2-1}\right) \geq \left(1 + \sqrt{2(x-1)}\right)$$

has been used.

In chapter III b we used the following estimate

$$\text{(A 14)} \quad | P_\ell(\cos\Theta_1) - P_\ell(\cos\Theta_2) | < C \frac{\sqrt{|\Theta_1 - \Theta_2|}}{(\sin\Theta_1 \sin\Theta_2)^{\frac{1}{4}}} \quad ; \quad 0 < \Theta_1, \Theta_2 < \pi$$

where C may be chosen independently of ℓ.

For the proof we first use (A 6) to obtain

$$\text{(A 15)} \quad | P_\ell(\cos\Theta_1) - P_\ell(\cos\Theta_2) | < 2\sqrt{\frac{1}{\pi(2\ell+1)}} \left\{ \frac{1}{(\sin\Theta_1)^{\frac{1}{2}}} + \frac{1}{(\sin\Theta_2)^{\frac{1}{2}}} \right\}.$$

Now we write

$$P_\ell(\cos\Theta_1) - P_\ell(\cos\Theta_2) = \int_{\cos\Theta_2}^{\cos\Theta_1} P_\ell'(\cos\Theta)\, d\cos\Theta$$

where we have

$$\text{(A 16)} \quad | P_\ell'(\cos\Theta) | < C \frac{\sqrt{(2\ell+1)}}{(\sin\Theta)^{\frac{3}{2}}} .$$

Therefore for Θ_1, Θ_2 in the interval $(0, \frac{2}{3}\pi)$, where $0.7\,\Theta < \sin\Theta < \Theta$

$$\text{(A 17)} \quad | P_\ell(\cos\Theta_1) - P_\ell(\cos\Theta_2) | < C\sqrt{2\ell+1}\, | \Theta_1^{\frac{1}{2}} - \Theta_2^{\frac{1}{2}} | .$$

Multiplying (A 15) and (A 17) we obtain

$$\text{(A 18)} \quad | P_\ell(\cos\Theta_1) - P_\ell(\cos\Theta_2) | < C \frac{\sqrt{|\Theta_1 - \Theta_2|}}{\Theta_1^{\frac{1}{4}}\, \Theta_2^{\frac{1}{4}}} .$$

Now we may replace (A 18) by

$$\text{(A 19)} \quad | P_\ell(\cos\Theta_1) - P_\ell(\cos\Theta_2) | < C \frac{\sqrt{|\Theta_1 - \Theta_2|}}{(\sin\Theta_1 \sin\Theta_2)^{\frac{1}{4}}}$$

which holds for $0 < \Theta_{1,2} < \pi$; $0 < \Theta_{2,1} < \frac{2}{3}\pi$ and

$\frac{\pi}{3} < \Theta_{1,2}$; $0 < \Theta_{2,1} < \pi$ where we used (A 5).

For the remaining case $0 < \Theta_{1,2} < \frac{\pi}{3}$; $\frac{2}{3}\pi < \Theta_{2,1} < \pi$ we use

$$| P_l(\cos\Theta_1) - P_l(\cos\Theta_2) | < 2$$

which proves (A 14).

We want to get some estimates on $P_l(z)$ if z becomes complex and varies in an ellipse with foci ± 1. Let $z = \cos(\Theta_1 + i\Theta_2)$ (Θ_1, Θ_2 real).

Then

$$\operatorname{Re} z = \cos\Theta_1 \cosh\Theta_2$$

$$\operatorname{Im} z = -\sin\Theta_1 \sinh\Theta_2$$

Therefore, if we choose Θ_2 to be constant, we just obtain the equation for an ellipse with foci ± 1 and semimajor axis $\cosh\Theta_2$:

$$\left(\frac{\operatorname{Re} z}{a}\right)^2 + \left(\frac{\operatorname{Im} z}{b}\right)^2 = 1$$

$$a = \cosh\Theta_2 , \quad b = \sinh\Theta_2 , \quad a^2 - b^2 = 1 .$$

Now define

$$\text{(A 20)} \qquad d^l_{\mu\nu}(\Theta) = \langle l\mu | \exp - i\Theta J_y | l\nu \rangle$$

where the $|l\mu\rangle$ $(-l \leq \mu \leq l)$ are the basis vectors of the representation space of $SU(2)$ with angular momentum l and where J_y is the representation of the generator of the rotations around the 2-axis.

Then in particular for integer l

$$\text{(A 21)} \qquad d^l_{00}(\Theta) = P_l(\cos\Theta)$$

so estimates on $d^l_{\mu\nu}$ will give estimates on P_l. Now we may write

$$d^l_{\mu\nu}(\Theta) = \sum_\lambda \langle l\mu | l, J_y = \lambda \rangle \langle l, J_y = \lambda | l\nu \rangle \, e^{-i\Theta\lambda}$$

which, when combined with Schwartz's inequality, leads to

$$| d^{\ell}_{\mu\lambda}(\Theta) | \leq \exp \ell\, \Theta_2 \sum_{\lambda} | \langle \ell\mu | \ell\, J_y = \lambda \rangle | \, | \langle \ell, J_y = \lambda | \ell\nu \rangle |$$

$$\leq \exp \ell\, \Theta_2$$

So if $\cos \Theta$ varies in an ellipse with foci ± 1 and semimajor axis we have

(A 22) $$| d^{\ell}_{\mu\nu}(\Theta) | \leq (x + \sqrt{x^2 - 1})^{\ell}$$

The combination of (A 13) and (A 22) proves the statement made on page 16.

Finally, using (A 4), it is easy to see that $| P_{\ell}(z) |$ takes its maximum at the right extremity of the ellipse. Generally this is seen to be true for $| d^{\ell}_{\lambda\lambda}(\Theta) |$ if ℓ is integer and for $| \cos \frac{\Theta}{2} \cdot d^{\ell}_{\lambda\lambda}(\Theta) |$ if ℓ is half-integer. This proves the statement made on page 36 and page 114 . Assume e.g. that

$$g(z) = \sum_{\ell} a_{\ell} P_{\ell}(z) \, , \quad a_{\ell} \geq 0$$

is a function analytic in an ellipse with foci ± 1 and semimajor axis $x > 1$ Then

$$| g(z) | \leq g(x)$$

for all z in that ellipse.

Literature

[A 1] Aks, J. J.M.P. 6 516 (1965)

[A 2] Ascoli, R. and A. Minguzzi, Phys. Rev. 118 1435 (1960)

[B 1] Bargmann, V. and E. Wigner, Proc. Nat. Acad. Sc. 34 211 (1948)

[B 2] Bell, J.S. CERN report TH 971 (1968)

[B 3] Bessis, J.D. and V. Glaser, N.C. 58 568 (1967)

[B 4] Blankenbecler, R. M.L. Goldberger, N.N. Khuri and S.B. Treiman, Ann.Phys. 10 62 (1960)

[B 5] Bochner, S. and W. Martin, Several complex variables, Princeton University Press, Princeton (1948).

[B 6] Bogoliubov, N.N., B.V. Medvedev and M.K. Polivanov, Voprossy Teorii Dispersionykh Sootnoshenii, Moscow (1958)

[B 7] Bonnier, R. and R. Vinh Mau, Phys. Rev. 165 1923 (1968)

[B 8] Bremermann, H.J.: Schriftenreihe Math. Inst. Univ. Münster 5 (1951)

[B 9] Bremermann, H.J., R. Oehme and J.G. Taylor, Phys. Rev. 109 2178 (1958)

[B 10] Bros, J., H. Epstein and V. Glaser, N.C. 31 1265 (1964)

[B 11] Bros, J., H. Epstein and V. Glaser, Comm. Math. Phys. 1, 240 (1965)

[C 1] Cerulus, F. and A. Martin, Phys. Lett. 8 80 (1963)

[C 2] Cheung, F.K. and J.S. Toll, Phys. Rev. 160 1072 (1967)

[C 3] Cohen-Tannoudji, A. Movel and H. Navelet, Ann. Phys. 46 239 (1968)

[C 4] Crichton, J.H., N.C. 45 A 256 (1966)

[D 1] Donnachie, A. and J. Hamilton, Phys. Rev. 133 1053 (1964)

[D 2] Dragt, A.J., Phys. Rev. 156 1588 (1967)

[E 1] Epstein, H., Axiomatic field theory, Brandeis university summer institute in theoretical physics, Gordon and Breach (1965)

[E 2] Epstein, H., V. Glaser and A. Martin, CERN report TH 991 (1969)

[E 3] Epstein, H., private communication

[E 4] Erdelyi, A. et al., The Bateman manuscript project, Mac Graw Hill, New York (1953)

[F 1] Froissart, M., Phys. Rev. 123 1053 (1961)

[G 1] Geshkenbein, B.V. and B.L. Ioffe, Soviet Phys. - JETP 17 820 (1963)

[G 2] Glaser, V. N.N. Bogoliubov's 60th anniversary memorial volume (to appear)

[G 3] Goldberger, M.K., Phys. Rev. 99 979 (1955)

[G 4] Greenberg, O.W. and F.E. Low, Phys. Rev. 124 2047 (1961)

[H 1] Hepp, K., Helv. Phys. Acta 37 639 (1964)
[H 2] Hörmander, L., Introduction to complex analysis in several variables, Van Nostrand, Princeton (1966)

[J 1] Jacob, M. and G.C. Wick, Ann. Phys. 7 404 (1959)
[J 2] Jin, Y.S. and A. Martin, Phys. Rev. 135 B 1369 (1964)
[J 3] Joos, H., Fortschr. Phys. 10 65 (1962)

[K 1] Källén, G., Elementary particle physics, Addison-Wesley, London (1964)
[K 2] Kantorovich, L.V. and G.P. Akilov, Functional analysis in normed spaces, Pergamon, New York (1964)
[K 3] Kinoshita, T., Loeffel, J.J. and A. Martin, Phys. Rev. Lett. 10 460 (1964), Phys. Rev. 135 B 1464 (1964)

[L 1] Lehmann, H., N.C. 10 579 (1958)
[L 2] Lehmann, H. Comm. Math. Phys. 2 375 (1966)
[L 3] Loginov, A.A., M.A. Mestrivishvili and N. van Hieu, Proceedings of the 1967 international conference on particles and fields, Interscience, New York (1967)
[L 4] Lukaszuk, L., N.C. 51 A 67 (1966)
[L 5] Lukaszuk, L. and A. Martin, N.C. 52 122 (1967)

[M 1] Mahoux, G., thèse, Université de Paris (1969)
[M 2] Mahoux, G. and A. Martin, Phys. Rev. 174 2140 (1968)
[M 3] Mandelstam, S., N.C. 15 658 (1960)
[M 4] Mandelstam, S., Phys. Rev. Lett. 4 84 (1960)
[M 5] Martin, A., Phys. Rev. 129 1432 (1963)
[M 6] Martin, A., N.C. 29 993 (1963)
[M 7] Martin, A., N.C. 39 704 (1965)
[M 8] Martin, A., High energy physics and elementary particles, I.A.E.A., Vienna (1965)
[M 9] Martin, A., N.C. 42 A 930 (1966)
[M 10] Martin, A., N.C. 44 1219 (1966)
[M 11] Martin, A., N.N. Bogoliubov's 60th anniversary memorial volume (to appear)
[M 12] Martin, A., N.C. 59 A 131 (1969)
[M 13] Meiman, N.N., Soviet Phys. JETP 17 830 (1963)
[M 14] Müller, V.F., N.C. 42 158 (1966)

[N 1] Nakanishi, N., Progr. Theor. Phys. 26 337 (1961)
[N 2] Newton, R.G., to appear in J.M.P.

[O 1] Odorice, R., N.C. 51 A 1021 (1967)

[R 1] Riahi, F., private communication

[S 1] Schiff, L.I., Quantum Mechanics, MacGraw Hill, New York (1955)

[S 2] Schwartz, L., Theorie des distributions, Hermann, Paris (1957)

[S 3] Smirnov, V.I., IZV. AN. SSSR, ser. matem. No. 3 (1932)

[S 4] Sommer, G., N.C. 48 A 92 (1967)

[S 5] Sommer, G., N.C. 52 A 373 (1967)

[S 6] Sommer, G., N.C. 52 A 850 (1967)

[S 7] Sommer, G., N.C. 52 A 866 (1967)

[S 8] Streater, R.F. and A.S. Wightman, Spin and Statistics and all that; Benjamin, New York (1964)

[S 9] Symanzik, K., Phys. Rev. 100 743 (1957)

[S 10] Szegö, G., Orthogonal Polynomials, American Math. Soc. Colloquium Publ., Vol. 23, New York (1959)

[T 1] Tiktopoulos, G. and S.B. Treiman, Phys. Rev. 167 1437 (1968)

[T 2] Titchmarsh, E.C., The theory of functions, Oxford University Press, Oxford (1939)

[T 3] Trueman, L.T. and G.C. Wick, Ann. Phys. 26 322 (1964)

[T 4] Trueman, L.T., Phys. Rev. Lett. 17 1198 (1966)

[V 1] Vladimirov, V.S., Methods of the theory of functions of many complex variables, M.I.T. Press, Cambridge (Mass.) (1966)

[W 1] Wang, L.L.C., Phys. Rev. 142 1187 (1966)

[W 2] Weinberg, S., Phys. Rev. Lett. 17 616 (1966)

[W 3] Wigner, E., Phys. Rev. 98 145 (1955)

[W 4] Williams, D., UCRL (1113)

[Y 1] Yamamoto, K., N.C. 27 1277 (1963)

[Y 2] Yosida, K., Functional analysis, Springer Verlag, Berlin, Göttingen, Heidelberg (1964)

[Z 1] Zimmermann, W., N.C. 21 249 (1961)

Lecture Notes in Physics

im Erscheinen / to appear

Vol. 1: J. C. Erdmann, Wärmeleitung in Kristallen, theoretische Grundlagen und fortgeschrittene experimentelle Methoden. 1969. DM 20,–

Vol. 2: K. Hepp, Théorie de la renormalisation. 1969. DM 18,–

Vol. 3: A. Martin, Scattering Theory: Unitarity, Analyticity and Crossing. 1969. DM 14,–

Selected Issues from

Lecture Notes in Mathematics

Vol. 7: Ph. Tondeur, Introduction to Lie Groups and Transformation Groups. VIII, 176 pages. 1965. DM 13,50

Vol. 8: G. Fichera, Linear Elliptic Differential Systems and Eigenvalue Problems. IV, 176 pages. 1965. DM 13,50

Vol. 17: C. Müller, Spherical Harmonics. IV, 46 pages. 1966. DM 5,–

Vol. 25: R. Narasimhan, Introduction to the Theory of Analytic Spaces. IV, 143 pages. 1966. DM 10,–

Vol. 33: G. I. Targonski, Seminar on Functional Operators and Equations. IV, 110 pages. 1967. DM 10,–

Vol. 35: N. P. Bhatia and G. P. Szegö, Dynamical Systems. Stability Theory and Applications. VI, 416 pages. 1967. DM 24,–

Vol. 40: J. Tits, Tabellen zu den einfachen Lie Gruppen und ihren Darstellungen. VI, 53 Seiten. 1967. DM 6,80

Vol. 45: A. Wilansky, Topics in Functional Analysis. VI, 102 pages. 1967. DM 9,60

Vol. 52: D. J. Simms, Lie Groups and Quantum Mechanics. IV, 90 pages. 1968. DM 8,–

Vol. 55: D. Gromoll, W. Klingenberg und W. Meyer, Riemannsche Geometrie im Großen. VI, 287 Seiten. 1968. DM 20,–

Vol. 56: K. Floret und J. Wloka, Einführung in die Theorie der lokalkonvexen Räume. VIII, 194 Seiten. 1968. DM 16,–

Vol. 60: Seminar on Differential Equations and Dynamical Systems. Edited by G. S. Jones. VI, 106 pages. 1968. DM 9,60

Vol. 81: J.-P. Eckmann et M. Guenin, Méthodes Algébriques en Mécanique Statistique. VI, 131 pages. 1969. DM 12,–

Vol. 82: J. Wloka, Grundräume und verallgemeinerte Funktionen. VIII, 131 Seiten. 1969. DM 12,–

Vol. 91: N. N. Janenko, Die Zwischenschrittmethode zur Lösung mehrdimensionaler Probleme der mathematischen Physik. VIII, 194 Seiten. 1969. DM 16,80

Vol. 102: F. Stummel, Rand- und Eigenwertaufgaben in Sobolewschen Räumen. VIII, 386 Seiten. 1969. DM 20,–

Vol. 103: Lectures in Modern Analysis and Applications I. Edited by C. T. Taam. VII, 162 pages. 1969. DM 12,–

Vol. 104: G. H. Pimbley, Jr., Eigenfunction Branches of Nonlinear Operators and their Bifurcations. II, 128 pages. 1969. DM 10,–

Beschaffenheit der Manuskripte

Die Manuskripte werden photomechanisch vervielfältigt; sie müssen daher in sauberer Schreibmaschinenschrift mit ausreichend großer Type geschrieben sein. Handschriftliche Formeln bitte nur mit schwarzer Tusche eintragen. Notwendige Korrekturen sind bei dem bereits geschriebenen Text entweder durch Überkleben des alten Textes vorzunehmen oder aber müssen die zu korrigierenden Stellen mit weißem Korrekturlack abgedeckt werden. Die reproduktionsfähigen Abbildungen (in Originalgröße) sollen in den Text eingeklebt werden. Falls das Manuskript oder Teile desselben neu geschrieben werden müssen, ist der Verlag bereit, dem Autor bei Erscheinen seines Bandes einen angemessenen Betrag zu zahlen. Die Autoren erhalten 50 Freiexemplare.

Zur Erreichung eines möglichst optimalen Reproduktionsergebnisses ist es erwünscht, daß bei der vorgesehenen Verkleinerung der Manuskripte der Text auf einer Seite in der Breite möglichst 18 cm und in der Höhe 26,5 cm nicht überschreitet. Entsprechende Satzspiegelvordrucke werden vom Verlag gern auf Anforderung zur Verfügung gestellt.

Manuskripte, in englischer, deutscher oder französischer Sprache abgefaßt, sind einzureichen bei: Springer-Verlag, 6900 Heidelberg, Postfach 1780.

Cette série a pour but de donner des informations rapides, de niveau élevé, sur des développements récents en physique, aussi bien dans la recherche que dans l'enseignement supérieur. On prévoit de publier.

1. des versions préliminaires de travaux originaux et de monographies
2. des cours spéciaux portant sur un domaine nouveau ou sur des aspects nouveaux de domaines classiques
3. des rapports de séminaires
4. des conférences faites lors de congrès ou de colloques

En outre il est prévu de publier dans cette série, si la demande le justifie, des rapports de séminaires et des cours multicopiés ailleurs mais déjà épuisés.

Dans l'intérêt d'une diffusion rapide, les contributions auront souvent un caractère provisoire; le cas échéant, les démonstrations ne seront données que dans les grandes lignes. Les travaux présentés pourront également paraître ailleurs. Une réserve suffisante d'exemplaires sera toujours disponible. En permettant aux personnes intéressées d'être informées plus rapidement, les éditeurs Springer espèrent, par cette série de «prépublications», rendre d'appréciables services aux instituts de physique. Les annonces dans les revues spécialisées, les inscriptions aux catalogues et les copyrights rendront plus facile aux bibliothèques la tâche de réunir une documentation complète.

Présentation des manuscrits

Les manuscrits, étant reproduits par procédé photomécanique, doivent être soigneusement dactylographiés (en caractère assez grand). Il est recommandé d'écrire à l'encre de Chine noire les formules non dactylographiées. Les corrections nécessaires doivent être effectuées soit par collage du nouveau texte sur l'ancien soit en recouvrant les endroits à corriger par du vernis correcteur blanc. Les illustrations; en dimension originale, préparées pour reproduction sont à insérer dans le texte. S'il s'avère nécessaire d'écrire de nouveau le manuscrit, soit complètement, soit en partie, la maison d'édition se déclare prête à verser à l'auteur, lors de la parution du volume, le montant des frais correspondants. Les auteurs recoivent 50 exemplaires gratuits.

Pour obtenir une reproduction optimale il est désirable que le texte dactylographié sur une page ne dépasse pas 26,5 cm en hauteur et 18 cm en largeur. Sur demande la maison d'edition met à la disposition des auteurs du papier spécialement préparé.

Les manuscrits en anglais, allemand ou français peuvent être adressés à Springer-Verlag, 6900 Heidelberg, Postfach 1780.